COMPLEX TOPOLOGICAL K-THEORY

Topological K-theory is a key tool in topology, differential geometry, and index theory, yet this is the first contemporary introduction for graduate students new to the subject. No background in algebraic topology is assumed; the reader need only have taken the standard first courses in real analysis, abstract algebra, and point-set topology.

The book begins with a detailed discussion of vector bundles and related algebraic notions, followed by the definition of K-theory and proofs of the most important theorems in the subject, such as the Bott periodicity theorem and the Thom isomorphism theorem. The multiplicative structure of K-theory and the Adams operations are also discussed and the final chapter details the construction and computation of characteristic classes.

With every important aspect of the topic covered, and exercises at the end of each chapter, this is the definitive book for a first course in topological K-theory.

CAMBRIDGE STUDIES IN ADVANCED MATHEMATICS

Editorial Board:

B. Bollobás, W. Fulton, A. Katok, F. Kirwan, P. Sarnak, B. Simon, B. Totaro

All the titles listed below can be obtained from good booksellers or from Cambridge University Press. For a complete series listing visit:
http://www.cambridge.org/series/sSeries.asp?code=CSAM

Already published

Complex Topological *K*-Theory

EFTON PARK

CAMBRIDGE
UNIVERSITY PRESS

CAMBRIDGE
UNIVERSITY PRESS

University Printing House, Cambridge CB2 8BS, United Kingdom

One Liberty Plaza, 20th Floor, New York, NY 10006, USA

477 Williamstown Road, Port Melbourne, VIC 3207, Australia

314-321, 3rd Floor, Plot 3, Splendor Forum, Jasola District Centre, New Delhi - 110025, India

103 Penang Road, #05-06/07, Visioncrest Commercial, Singapore 238467

Cambridge University Press is part of the University of Cambridge.

It furthers the University's mission by disseminating knowledge in the pursuit of
education, learning and research at the highest international levels of excellence.

www.cambridge.org
Information on this title: www.cambridge.org/9780521856348

First published 2008

A catalogue record for this publication is available from the British Library

ISBN 978-0-521-85634-8 Hardback

To Alex, Connor, Nolan, and Rhonda

Contents

Preface

Topological K-theory first appeared in a 1961 paper by Atiyah and Hirzebruch; their paper adapted the work of Grothendieck on algebraic varieties to a topological setting. Since that time, topological K-theory (which we will henceforth simply call K-theory) has become a powerful and indespensible tool in topology, differential geometry, and index theory. The goal of this book is to provide a self-contained introduction to the subject.

This book is primarily aimed at beginning graduate students, but also for working mathematicians who know little or nothing about the subject and would like to learn something about it. The material in this book is suitable for a one semester course on K-theory; for this reason, I have included exercises at the end of each chapter. I have tried to keep the prerequisites for reading this book to a minimum; I will assume that the reader knows the following:

- Linear Algebra: Vector spaces, bases, linear transformations, similarity, trace, determinant.

- Abstract Algebra: Groups, rings, homomorphisms and isomorphisms, quotients, products.

- Topology: Metric spaces, completeness, compactness and connectedness, local compactness, continuous functions, quotient topology, subspace topology, partitions of unity.

To appreciate many of the motivating ideas and examples in K-theory, it is helpful, but not essential, for the reader to know the rudiments of differential topology, such as smooth manifolds, tangent bundles, differential forms, and de Rham cohomology. In Chapter 4, the theory of characteristic classes is developed in terms of differential forms and de

Rham cohomology; for readers not familiar with these topics, I give a quick introduction at the beginning of that chapter. I do not assume that the reader has any familiarity with homological algebra; the necessary ideas from this subject are developed at the end of Chapter 1.

To keep this book short and as easy to read as possible (especially for readers early in their mathematical careers), I have kept the scope of this book very limited. Only complex K-theory is discussed, and I do not say anything about equivariant K-theory. I hope the reader of this book will be inspired to learn about other versions of K-theory; see the bibliography for suggestions for further reading.

It is perhaps helpful to say a little bit about the philosophy of this book, and how this book differs from other books on K-theory. The fundamental objects of study in K-theory are vector bundles over topological spaces (in the case of K^0) and automorphisms of vector bundles (in the case of K^1). These concepts are discussed at great length in this book, but most of the proofs are formulated in terms of the equivalent notions of idempotents and invertible matrices over Banach algebras of continuous complex-valued functions. This more algebraic approach to K-theory makes the presentation "cleaner"(in my opinion), and also allows readers to see how K-theory can be extended to matrices over general Banach algebras. Because commutativity of the Banach algebras is not necessary to develop K-theory, this generalization falls into an area of mathematics that is often referred to as *noncommutative topology*. On the other hand, there are important aspects of K-theory, such as the existence of operations and multiplicative structures, that do not carry over to the noncommutative setting, and so we will restrict our attention to the K-theory of topological spaces.

I thank my colleagues, friends, and family for their encouragement while I was writing this book, and I especially thank Scott Nollet and Greg Friedman for reading portions of the manuscript and giving me many helpful and constructive suggestions.

1
Preliminaries

The goal of K-theory is to study and understand a topological space X by associating to it a sequence of abelian groups. The algebraic properties of these groups reflect topological properties of X, and the overarching philosophy of K-theory (and, indeed, of all algebraic topology) is that we can usually distinguish groups more easily than we can distinguish topological spaces. There are many variations on this theme, such as homology and cohomology groups of various sorts. What sets K-theory apart from its algebraic topological brethren is that not only can it be defined directly from X, but also in terms of matrices of continuous complex-valued functions on X. For this reason, we devote a significant part of this chapter to the study of matrices of continuous functions.

Our first step is to look at complex vector spaces equipped with an inner product. The reader is presumably familiar with inner products on real vector spaces, but possibly not the complex case. For this reason, we begin with a brief discussion of complex inner product spaces.

1.1 Complex inner product spaces

Definition 1.1.1 *Let $\mathcal{V}$ be a finite-dimensional complex vector space and let $\mathbb{C}$ denote the complex numbers. A (complex) inner product on $\mathcal{V}$ is a function $\langle \cdot, \cdot \rangle : \mathcal{V} \times \mathcal{V} \longrightarrow \mathbb{C}$ such that for all elements v, v', and v'' in $\mathcal{V}$ and all complex numbers α and β:*

(i) $\langle \alpha v + \beta v', v'' \rangle = \alpha \langle v, v'' \rangle + \beta \langle v', v'' \rangle$;

(ii) $\langle v, \alpha v' + \beta v'' \rangle = \overline{\alpha} \langle v, v' \rangle + \overline{\beta} \langle v, v'' \rangle$;

(iii) $\langle v', v \rangle = \overline{\langle v, v' \rangle}$;

(iv) $\langle v, v \rangle \geq 0$, *with* $\langle v, v \rangle = 0$ *if and only if* $v = 0$.

1

For each v in $\mathcal{V}$, the nonnegative number $\|v\|_{in} = \sqrt{\langle v, v \rangle}$ is called the magnitude of v. A vector space equipped with an inner product is called a (complex) inner product space. A vector space basis $\{v_1, v_2, \ldots, v_n\}$ of $\mathcal{V}$ is orthogonal *if $\langle v_j, v_k \rangle = 0$ for $j \neq k$, and* orthonormal *if it is orthogonal and $\|v_k\|_{in} = 1$ for all $1 \leq k \leq n$.*

Proposition 1.1.2 *Every complex inner product space $\mathcal{V}$ admits an orthonormal basis.*

Proof The proof of this proposition follows the same lines as the corresponding fact for real inner product spaces. Start with any vector space basis $\{v_1, v_2, \ldots, v_n\}$ of $\mathcal{V}$ and apply the Gram–Schmidt process inductively to define an orthogonal basis

$$v_1' = v_1$$

$$v_2' = v_2 - \frac{\langle v_2, v_1' \rangle}{\langle v_1', v_1' \rangle} v_1'$$

$$\vdots$$

$$v_n' = v_n - \frac{\langle v_n, v_1' \rangle}{\langle v_1', v_1' \rangle} v_1' - \frac{\langle v_n, v_2' \rangle}{\langle v_2', v_2' \rangle} v_2' - \cdots - \frac{\langle v_n, v_{n-1}' \rangle}{\langle v_{n-1}', v_{n-1}' \rangle} v_{n-1}'.$$

Then

$$\left\{ \frac{v_1'}{\|v_1'\|_{in}}, \frac{v_2'}{\|v_2'\|_{in}}, \cdots, \frac{v_n'}{\|v_n'\|_{in}} \right\}$$

is an orthonormal basis of $\mathcal{V}$. □

For elements $(z_1, z_2, \ldots, z_n)$ and $(z_1', z_2', \ldots, z_n')$ in the vector space $\mathbb{C}^n$, the formula

$$\langle (z_1, z_2, \ldots, z_n), (z_1', z_2', \ldots, z_n') \rangle = z_1 \overline{z}_1' + z_2 \overline{z}_2' + \cdots + z_n \overline{z}_n'$$

defines the *standard inner product* on $\mathbb{C}^n$. For each $1 \leq k \leq n$, define e_k to be the vector that is 1 in the kth component and 0 elsewhere. Then $\{e_1, e_2, \cdots, e_n\}$ is the *standard orthonormal basis* for $\mathbb{C}^n$.

Proposition 1.1.3 (Cauchy–Schwarz inequality) *Let $\mathcal{V}$ be an inner product space. Then*

$$|\langle v, v' \rangle| \leq \|v\|_{in} \|v'\|_{in}$$

for all v and v' in $\mathcal{V}$.

Proof If $\langle v, v' \rangle = 0$, the proposition is trivially true, so suppose that $\langle v, v' \rangle \neq 0$. For any α in $\mathbb{C}$, we have

$$
\begin{aligned}
0 \leq \|\alpha v + v'\|_{in}^2 &= \langle \alpha v + v', \alpha v + v' \rangle \\
&= |\alpha|^2 \|v\|_{in}^2 + \|v'\|_{in}^2 + \alpha \langle v, v' \rangle + \overline{\alpha} \overline{\langle v, v' \rangle} \\
&= |\alpha|^2 \|v\|_{in}^2 + \|v'\|_{in}^2 + 2 \operatorname{Re}(\alpha \langle v, v' \rangle),
\end{aligned}
$$

where $\operatorname{Re}(\alpha \langle v, v' \rangle)$ denotes the real part of $\alpha \langle v, v' \rangle$. Take α to have the form $t \overline{\langle v, v' \rangle} | \langle v, v' \rangle |^{-1}$ for t real. Then the string of equalities above yields

$$
\|v\|_{in}^2 t^2 + 2| \langle v, v' \rangle | t + \|v'\|_{in}^2 \geq 0
$$

for all real numbers t. This quadratic equation in t has at most one real root, implying that

$$
4| \langle v, v' \rangle |^2 - 4 \|v\|_{in}^2 \|v'\|_{in}^2 \leq 0,
$$

whence the proposition follows. $\qquad\square$

Proposition 1.1.4 (Triangle inequality) *Let $\mathcal{V}$ be an inner product space. Then*

$$
\|v + v'\|_{in} \leq \|v\|_{in} + \|v'\|_{in}
$$

for all v and v' in $\mathcal{V}$.

Proof Proposition 1.1.3 gives us

$$
\begin{aligned}
\|v + v'\|_{in}^2 &= \langle v + v', v + v' \rangle \\
&= \langle v, v \rangle + \langle v, v' \rangle + \langle v', v \rangle + \langle v', v' \rangle \\
&= \|v\|_{in}^2 + \|v'\|_{in}^2 + 2 \operatorname{Re} \langle v, v' \rangle \\
&\leq \|v\|_{in}^2 + \|v'\|_{in}^2 + 2| \langle v, v' \rangle | \\
&\leq \|v\|_{in}^2 + \|v'\|_{in}^2 + 2 \|v\|_{in} \|v'\|_{in} \\
&= (\|v\|_{in} + \|v'\|_{in})^2 .
\end{aligned}
$$

We get the desired result by taking square roots. $\qquad\square$

Definition 1.1.5 *Let $\mathcal{V}$ be an inner product space and let $\mathcal{W}$ be a vector subspace of $\mathcal{V}$. The vector subspace*

$$
\mathcal{W}^\perp = \{ v \in \mathcal{V} : \langle v, w \rangle = 0 \text{ for all } w \in \mathcal{W} \}
$$

is called the orthogonal complement *of $\mathcal{W}$ in $\mathcal{V}$.*

Proposition 1.1.6 *Let $\mathcal{V}$ be an inner product space and suppose that $\mathcal{W}$ is a vector subspace of $\mathcal{V}$. Then $\mathcal{V} \cong \mathcal{W} \oplus \mathcal{W}^{\perp}$.*

Proof If u is in the intersection of $\mathcal{W}$ and $\mathcal{W}^{\perp}$, then $\|u\|_{in} = \langle u, u \rangle = 0$, whence $u = 0$. Take v in $\mathcal{V}$, and suppose that $v = w_1 + w_1^{\perp} = w_2 + w_2^{\perp}$ for w_1, w_2 in $\mathcal{W}$ and $w_1^{\perp}, w_2^{\perp}$ in $\mathcal{W}^{\perp}$. Then $w_1 - w_2 = w_2^{\perp} - w_1^{\perp}$ is in $\mathcal{W} \cap \mathcal{W}^{\perp}$ and therefore we must have $w_1 = w_2$ and $w_1^{\perp} = w_2^{\perp}$. To show that such a decomposition of v actually exists, choose an orthonormal basis $\{w_1, w_2, \ldots, w_m\}$ of $\mathcal{W}$ and set $w = \sum_{k=1}^{m} \langle v, w_k \rangle w_k$. Clearly w is in $\mathcal{W}$. Moreover, for every $1 \leq j \leq m$, we have

$$\langle v - w, w_j \rangle = \langle v, w_j \rangle - \langle w, w_j \rangle$$
$$= \langle v, w_j \rangle - \sum_{k=1}^{m} \langle v, w_k \rangle \langle w_k, w_j \rangle$$
$$= \langle v, w_j \rangle - \langle v, w_j \rangle = 0,$$

which implies that $v - w$ is in $\mathcal{W}^{\perp}$. $\square$

Definition 1.1.7 *Let $\mathcal{V}$ be an inner product space, let $\mathcal{W}$ be a vector subspace of $\mathcal{V}$, and identify $\mathcal{V}$ with $\mathcal{W} \oplus \mathcal{W}^{\perp}$. The linear map $\mathsf{P} : \mathcal{V} \longrightarrow \mathcal{W}$ given by $\mathsf{P}(w, w^{\perp}) = w$ is called the* orthogonal projection *of $\mathcal{V}$ onto $\mathcal{W}$.*

We close this section with a notion that we will need in Chapter 3.

Proposition 1.1.8 *Let $\mathcal{V}$ and $\mathcal{W}$ be inner product spaces, and suppose that $\mathsf{A} : \mathcal{V} \longrightarrow \mathcal{W}$ is a vector space homomorphism; i.e., a linear map. Then there exists a unique vector space homomorphism $\mathsf{A}^* : \mathcal{W} \longrightarrow \mathcal{V}$, called the* adjoint *of A, for which $\langle \mathsf{A}v, w \rangle = \langle v, \mathsf{A}^*w \rangle$ for all v in $\mathcal{V}$ and w in $\mathcal{W}$.*

Proof Fix orthonormal bases $\{e_1, e_2, \ldots, e_m\}$ and $\{f_1, f_2, \ldots, f_n\}$ for $\mathcal{V}$ and $\mathcal{W}$ respectively. For each $1 \leq i \leq m$, write $\mathsf{A}e_i$ in the form $\mathsf{A}e_i = \sum_{j=1}^{n} a_{ji} f_j$ and set $\mathsf{A}^* f_j = \sum_{i=1}^{m} \bar{a}_{ji} e_i$ Then

$$\langle \mathsf{A}e_i, f_j \rangle = a_{ji} = \langle e_i, \mathsf{A}^* f_j \rangle$$

for all i and j, and parts (i) and (ii) of Definition 1.1.1 imply that $\langle \mathsf{A}v, w \rangle = \langle v, \mathsf{A}^*w \rangle$ for all v in $\mathcal{V}$ and w in $\mathcal{W}$.

To show uniqueness, suppose that $B : \mathcal{W} \longrightarrow \mathcal{V}$ is a linear map with the property that $\langle \mathsf{A}v, w \rangle = \langle v, \mathsf{A}^*w \rangle = \langle v, Bw \rangle$ for all v and w. Then $\langle v, (\mathsf{A}^* - B)w \rangle = 0$, and by taking $v = (\mathsf{A}^* - B)w$ we see that $(\mathsf{A}^* - B)w = 0$ for all w. Thus $\mathsf{A}^* = B$.

To prove that A^* is a vector space homomorphism, note that

$$\langle v, A^*(\alpha w + \beta w') \rangle = \langle Av, \alpha w + \beta w' \rangle$$
$$= \overline{\alpha} \langle Av, w \rangle + \overline{\beta} \langle Av, w' \rangle$$
$$= \overline{\alpha} \langle v, A^*w \rangle + \overline{\beta} \langle v, A^*w' \rangle$$
$$= \langle v, \alpha A^*w + \beta A^*w' \rangle$$

for all v in $\mathcal{V}$, all w and w' in $\mathcal{W}$, and all complex numbers α and β. Therefore $A^*(\alpha w + \beta w') = \alpha A^*w + \beta A^*w'$. □

Proposition 1.1.9 *Let $\mathcal{U}$, $\mathcal{V}$, and $\mathcal{W}$ be inner product spaces, and suppose that $A : \mathcal{U} \longrightarrow \mathcal{V}$ and $B : \mathcal{V} \longrightarrow \mathcal{W}$ are vector space homomorphisms. Then:*

(i) $(A^*)^* = A$;
(ii) $A^*B^* = BA^*$;
(iii) A^* *is an isomorphism if and only if A is an isomorphism.*

Proof The uniqueness of the adjoint and the equalities

$$\langle A^*v, u \rangle = \overline{\langle u, A^*v \rangle} = \overline{\langle Av, u \rangle} = \langle v, Au \rangle$$

for all u in $\mathcal{U}$ and v in $\mathcal{V}$ give us (i), and the fact that

$$\langle BAu, w \rangle = \langle Au, B^*w \rangle = \langle u, A^*B^*w \rangle$$

for all u in $\mathcal{U}$ and w in $\mathcal{W}$ establishes (ii).

If A is an isomorphism, then $\mathcal{U}$ and $\mathcal{V}$ have the same dimension and thus we can show A^* is an isomorphism by showing that A^* is injective. Suppose that $A^*v = 0$. Then $0 = \langle u, A^*v \rangle = \langle Au, v \rangle$ for all u in $\mathcal{U}$. But A is surjective, so $\langle v, v \rangle = 0$, whence $v = 0$ and A^* is injective. The reverse implication in (iii) follows from replacing A by A^* and invoking (i). □

1.2 Matrices of continuous functions

Definition 1.2.1 *Let X be a compact Hausdorff space. The set of all complex-valued continuous functions on X is denoted $C(X)$. If m and n are natural numbers, the set of m by n matrices with entries in $C(X)$ is written $M(m, n, C(X))$. If $m = n$, we shorten $M(m, n, C(X))$ to $M(n, C(X))$.*

Each of these sets of matrices has the structure of a *Banach space*:

Definition 1.2.2 *A Banach space is a vector space $\mathcal{V}$ equipped with a function $\|\cdot\| : \mathcal{V} \longrightarrow [0, \infty)$, called a* norm, *satisfying the following properties:*

   (i) *For all v and v' in $\mathcal{V}$ and α in $\mathbb{C}$:*

      (a) $\|\alpha v\| = |\alpha| \, \|v\|$;

      (b) $\|v + v'\| \le \|v\| + \|v'\|$.

   (ii) *The formula $d(v, v') = \|v - v'\|$ is a distance function on $\mathcal{V}$ and $\mathcal{V}$ is complete with respect to d.*

The topology generated by $d(v, w) = \|v - w\|$ is called the *norm topology* on $\mathcal{V}$; an easy consequence of the axioms is that scalar multiplication and addition are continuous operations in the norm topology.

Note that when X is a point we can identify $C(X)$ with $\mathbb{C}$.

Lemma 1.2.3 *For all natural numbers m and n, the set of matrices $\mathrm{M}(m, n, \mathbb{C})$ is a Banach space in the operator norm*

$$\|A\|_{op} = \sup \left\{ \frac{\|A\vec{z}\|_{in}}{\|\vec{z}\|_{in}} : \vec{z} \in \mathbb{C}^n, \vec{z} \ne 0 \right\}$$
$$= \sup \left\{ \|A\vec{z}\|_{in} : \|\vec{z}\|_{in} = 1 \right\}.$$

Proof For each A in $\mathrm{M}(m, n, \mathbb{C})$, we have

$$\|A\|_{op} = \sup \left\{ \frac{\|A\vec{w}\|_{in}}{\|\vec{w}\|_{in}} : \vec{w} \ne 0 \right\}$$
$$= \sup \left\{ \left\| A\left(\frac{\vec{w}}{\|\vec{w}\|_{in}} \right) \right\|_{in} : \vec{w} \ne 0 \right\}$$
$$= \sup \{ \|A\vec{z}\|_{in} : \|\vec{z}\|_{in} = 1 \},$$

and thus the two formulas for the operator norm agree. The equation $\|A(\lambda\vec{z})\|_{in} = |\lambda| \, \|A\vec{z}\|_{in}$ yields $\|\lambda A\|_{op} = |\lambda| \, \|A\|_{op}$, and the inequality $\|A_1 + A_2\|_{op} \le \|A_1\|_{op} + \|A_2\|_{op}$ is a consequence of Proposition 1.1.4.

To show completeness, let $\{A_k\}$ be a Cauchy sequence in $\mathrm{M}(m, n, \mathbb{C})$. Then for each $\vec{z}$ in $\mathbb{C}^n$, the sequence $\{A_k\vec{z}\}$ in $\mathbb{C}^m$ is Cauchy and therefore has a limit. Continuity of addition and scalar multiplication imply that the function $\vec{z} \mapsto \lim_{k\to\infty} A_k\vec{z}$ defines a linear map from $\mathbb{C}^n$ to $\mathbb{C}^m$. Take the standard vector space bases of $\mathbb{C}^m$ and $\mathbb{C}^n$ and let A denote the corresponding matrix in $\mathrm{M}(m, n, \mathbb{C})$; we must show that $\{A_k\}$ converges in norm to A.

Fix $\epsilon > 0$ and choose a natural number N with the property that $\|A_k - A_l\|_{op} < \epsilon/2$ for $k, l > N$. Then

$$\|A_k \vec{z} - A\vec{z}\|_{in} = \lim_{l \to \infty} \|A_k \vec{z} - A_l \vec{z}\|_{in}$$
$$\leq \limsup_{l \to \infty} \|A_k - A_l\|_{op} \|\vec{z}\|_{in}$$
$$< \epsilon \|\vec{z}\|_{in}$$

for all $\vec{z} \neq 0$ in $\mathbb{C}^n$. Hence $\|A_k - A\|_{op} < \epsilon$ for $k > N$, and the desired conclusion follows. $\qquad\square$

For the case where $m = n = 1$, the norm on each z in $M(1, \mathbb{C}) = \mathbb{C}$ defined in Lemma 1.2.3 is simply the modulus $|z|$.

Proposition 1.2.4 *Let X be a compact Hausdorff space and let m and n be natural numbers. Then $M(m, n, C(X))$ is a Banach space in the supremum norm*

$$\|A\|_{\infty} = \sup\{\|A(x)\|_{op} : x \in X\}.$$

Proof The operations of pointwise matrix addition and scalar multiplication make $M(m, n, C(X))$ into a vector space. Note that

$$\|\alpha A\|_{\infty} = \sup\{\|\alpha A(x)\|_{op} : x \in X\}$$
$$= \sup\{|\alpha| \, \|A(x)\|_{op} : x \in X\} = |\alpha| \, \|A\|_{\infty}$$

and

$$\|A + B\|_{\infty} = \sup\{\|A(x) + B(x)\|_{op} : x \in X\}$$
$$\leq \sup\{\|A(x)\|_{op} : x \in X\} + \sup\{\|B(x)\|_{op} : x \in X\}$$
$$= \|A\|_{\infty} + \|B\|_{\infty}$$

for all A and B in $M(m, n, C(X))$ and α in $\mathbb{C}$, and thus $\|\cdot\|_{\infty}$ is indeed a norm.

To check that $M(m, n, C(X))$ is complete in the supremum norm, let $\{A_k\}$ be a Cauchy sequence in $M(m, n, C(X))$. For each x in X, the sequence $\{A_k(x)\}$ is a Cauchy sequence in $M(m, n, \mathbb{C})$ and therefore by Lemma 1.2.3 has a limit $A(x)$. To show that this construction yields an element A in $M(m, n, C(X))$, we need to show that the (i, j) entry A_{ij} of A is in $C(X)$ for all $1 \leq i \leq m$ and $1 \leq j \leq n$.

Fix i and j. To simplify notation, let $f = A_{ij}$, and for each natural number k, let $f_k = (A_k)_{ij}$; note that each f_k is an element of $C(X) =$

$M(1, C(X))$. Endow $\mathbb{C}^m$ and $\mathbb{C}^n$ with their standard orthonormal bases. For each x in X, we have $f_k(x) = \langle \mathsf{A}_k(x)e_j, e_i \rangle$ and $f(x) = \langle \mathsf{A}(x)e_j, e_i \rangle$. Then for all natural numbers k and l, Proposition 1.1.3 gives us

$$
\begin{aligned}
|f_k(x) - f_l(x)| &= |\langle \mathsf{A}_k e_j, e_i \rangle - \langle \mathsf{A}_l e_j, e_i \rangle| \\
&= |\langle (\mathsf{A}_k - \mathsf{A}_l)e_j, e_i \rangle| \\
&\leq \|(\mathsf{A}_k - \mathsf{A}_l)e_j\|_{in} \, \|e_i\|_{in} \\
&\leq \|\mathsf{A}_k - \mathsf{A}_l\|_{op} \, \|e_j\|_{in} \, \|e_i\|_{in} \\
&= \|\mathsf{A}_k - \mathsf{A}_l\|_{op} \, .
\end{aligned}
$$

Therefore $\{f_k(x)\}$ is Cauchy and thus converges to $f(x)$.

To show that f is continuous, fix $\epsilon > 0$ and choose a natural number M with the property that $\|f_k - f_M\|_\infty < \epsilon/3$ for all $k > M$. Next, choose x' in X and let U be an open neighborhood of x' with the property that $|f_M(x') - f_M(x)| < \epsilon/3$ for all x in U. Then

$$
\begin{aligned}
|f(x') - f(x)| &\leq |f(x') - f_M(x')| + |f_M(x') - f_M(x)| + |f_M(x) - f(x)| \\
&< \lim_{k \to \infty} |f_k(x') - f_M(x')| + \frac{\epsilon}{3} + \lim_{k \to \infty} |f_M(x) - f_k(x)| \\
&\leq \limsup_{k \to \infty} \|f_k - f_M\|_\infty + \frac{\epsilon}{3} + \limsup_{k \to \infty} \|f_M - f_k\|_\infty \\
&< \epsilon
\end{aligned}
$$

for all $k > M$ and x in U, whence f is continuous.

The last step is to show that the sequence $\{\mathsf{A}_k\}$ converges in the supremum norm to A. Fix $\epsilon > 0$ and choose a natural number N so large that $\|\mathsf{A}_k - \mathsf{A}_l\|_\infty < \epsilon/2$ whenever k and l are greater than N. Then

$$
\begin{aligned}
\|\mathsf{A}_k(x) - \mathsf{A}(x)\|_{op} &= \lim_{l \to \infty} \|\mathsf{A}_k(x) - \mathsf{A}_l(x)\|_{op} \\
&\leq \limsup_{l \to \infty} \|\mathsf{A}_k - \mathsf{A}_l\|_\infty \\
&\leq \frac{\epsilon}{2} < \epsilon
\end{aligned}
$$

for $k > N$ and x in X. This inequality holds for each x in X and therefore

$$
\lim_{k \to \infty} \|\mathsf{A}_k - \mathsf{A}\|_\infty = 0.
$$

$\square$

In this book we will work almost exclusively with square matrices.

This will allow us to endow our Banach spaces $M(n, C(X))$ with an additional operation that gives us an *algebra*:

Definition 1.2.5 *An* algebra *is a vector space $\mathcal{V}$ equipped with a multiplication $\mathcal{V} \times \mathcal{V} \longrightarrow \mathcal{V}$ that makes $\mathcal{V}$ into a ring, possibly without unit, and satisfies $\alpha(vv') = (\alpha v)v' = v(\alpha v')$ for all v and v' in $\mathcal{V}$ and α in $\mathbb{C}$. If in addition $\mathcal{V}$ is a Banach space such that $\|vv'\| \leq \|v\| \, \|v'\|$ for all v and v' in $\mathcal{V}$, we call $\mathcal{V}$ a* Banach algebra.

Proposition 1.2.6 *Let X be a compact Hausdorff space and let n be a natural number. Then $M(n, C(X))$ is a Banach algebra with unit under matrix multiplication.*

Proof Proposition 1.2.4 tells us that $M(n, C(X))$ is a Banach space, and the reader can check that $M(n, C(X))$ is an algebra under pointwise matrix multiplication. To complete the proof, observe that

$$\begin{aligned}
\|AB\|_\infty &= \sup\{\|A(x)B(x)\|_{op} : x \in X\} \\
&\leq \sup\{\|A(x)\|_{op} : x \in X\} \sup\{\|B(x)\|_{op} : x \in X\} \\
&= \|A\|_\infty \|B\|_\infty
\end{aligned}$$

for all A and B in $M(n, C(X))$. $\qquad\qquad\square$

Before we leave this section, we establish some notation. We will write the zero matrix and the identity matrix in $M(n, C(X))$ as 0_n and I_n respectively when we want to highlight the matrix size. Next, suppose that B is an element of $M(n, C(X))$ and that A is a subspace of X. Then B restricts to define an element of $M(n, C(A))$; we will use the notation $B|A$ for this restricted matrix.

Finally, we will often be working with matrices that have block diagonal form, and it will be convenient to have a more compact notation for such matrices. Given matrices A and B in $M(m, C(X))$ and $M(n, C(X))$ respectively we set

$$\mathrm{diag}(A, B) = \begin{pmatrix} A & 0 \\ 0 & B \end{pmatrix} \in M(m + n, C(X)).$$

We will use the obvious extension of this notation for matrices that are comprised of more than two blocks.

1.3 Invertibles

Invertible matrices play several important roles in defining K-theory groups of a topological space. In this section we will prove various results about such matrices.

Definition 1.3.1 *Let X be compact Hausdorff. For each natural number n, the group of invertible elements of $\mathrm{M}(n, C(X))$ under multiplication is denoted $\mathrm{GL}(n, C(X))$.*

We begin by defining an important family of invertible matrices.

Definition 1.3.2 *Let n be a natural number. For every $0 \leq t \leq 1$, define the matrix*

$$\mathsf{Rot}(t) = \begin{pmatrix} \cos(\frac{\pi t}{2})I_n & -\sin(\frac{\pi t}{2})I_n \\ \sin(\frac{\pi t}{2})I_n & \cos(\frac{\pi t}{2})I_n \end{pmatrix}.$$

Note that for each t, the matrix $\mathsf{Rot}(t)$ is invertible with inverse

$$\mathsf{Rot}^{-1}(t) = \begin{pmatrix} \cos(\frac{\pi t}{2})I_n & \sin(\frac{\pi t}{2})I_n \\ -\sin(\frac{\pi t}{2})I_n & \cos(\frac{\pi t}{2})I_n \end{pmatrix}.$$

Proposition 1.3.3 *Let X be a compact Hausdorff space, let n be a natural number, and suppose S and T are elements of $\mathrm{GL}(n, C(X))$. Then*

$$\mathrm{diag}(\mathsf{S}, I_n)\mathsf{Rot}(t)\,\mathrm{diag}(\mathsf{T}, I_n)\mathsf{Rot}^{-1}(t)$$

is a homotopy in $\mathrm{GL}(2n, C(X))$ from $\mathrm{diag}(\mathsf{ST}, I_n)$ to $\mathrm{diag}(\mathsf{S}, \mathsf{T})$.

Proof Compute.   □

Proposition 1.3.4 *Let X be a compact Hausdorff space, let n be a natural number, and suppose that S in $\mathrm{M}(n, C(X))$ has the property that $\|I_n - \mathsf{S}\|_\infty < 1$. Then S is in $\mathrm{GL}(n, C(X))$ and*

$$\|\mathsf{S}^{-1}\|_\infty \leq \frac{1}{1 - \|I_n - \mathsf{S}\|_\infty}.$$

Proof Because

$$\left\| \sum_{j=0}^{k}(I_n - S)^j - \sum_{j=0}^{N}(I_n - S)^j \right\|_\infty = \left\| \sum_{j=N+1}^{k}(I_n - S)^j \right\|_\infty$$

$$\leq \sum_{j=N+1}^{k} \|I_n - S\|_\infty^j$$

$$\leq \sum_{j=N+1}^{\infty} \|I_n - S\|_\infty^j$$

$$= \frac{\|I_n - S\|_\infty^{N+1}}{1 - \|I_n - S\|_\infty}$$

for natural numbers $k > N$, the partial sums $\{\sum_{j=0}^{k}(I_n - S)^j\}$ form a Cauchy sequence. Let T be its limit. Then

$$ST = \lim_{k\to\infty} S \sum_{j=0}^{k}(I_n - S)^j$$

$$= \lim_{k\to\infty} [I_n - (I_n - S)] \sum_{j=0}^{k}(I_n - S)^j$$

$$= \lim_{k\to\infty} I_n - (I_n - S)^{k+1}$$

$$= I_n.$$

A similar argument yields $TS = I_n$, and so $T = S^{-1}$. Finally, the computation

$$\left\| S^{-1} \right\|_\infty = \lim_{k\to\infty} \left\| \sum_{j=0}^{k}(I_n - S)^j \right\|_\infty$$

$$\leq \lim_{k\to\infty} \sum_{j=0}^{k} \|I_n - S\|_\infty^j$$

$$= \frac{1}{1 - \|I_n - S\|_\infty}$$

establishes the final statement of the lemma. $\qquad\square$

Proposition 1.3.5 *Let X be a compact Hausdorff space, let n be a natural number, and choose S in $\mathrm{GL}(n, C(X))$. Suppose that $T \in \mathrm{M}(n, C(X))$*

has the property that

$$\|S - T\|_\infty < \frac{1}{\|S^{-1}\|_\infty}.$$

Then $tS + (1 - t)T$ *is in* $\mathrm{GL}(n, C(X))$ *for all* $0 \le t \le 1$. *In particular,* T *is in* $\mathrm{GL}(n, C(X))$.

Proof For all $0 \le t \le 1$, we have

$$
\begin{aligned}
1 &> \|S^{-1}\|_\infty \|S - T\|_\infty \\
&\ge \|I_n - S^{-1}T\|_\infty \\
&\ge \|(1 - t)(I_n - S^{-1}T)\|_\infty \\
&= \|(I_n - (tI_n + (1 - t)S^{-1}T))\|_\infty .
\end{aligned}
$$

By Proposition 1.3.4, the matrix $tI_n + (1 - t)S^{-1}T$ is invertible. The product of invertible elements is invertible and thus

$$S(tI_n + (1 - t)S^{-1}T) = tS + (1 - t)T$$

is invertible. □

Corollary 1.3.6 *Let* X *be compact Hausdorff. For every natural number* n, *the set* $\mathrm{GL}(n, C(X))$ *is open in* $\mathrm{M}(n, C(X))$.

Proof Proposition 1.3.5 shows that for each S in $\mathrm{GL}(n, C(X))$, the elements of $\mathrm{M}(n, C(X))$ in the open ball of radius $1/\|S^{-1}\|_\infty$ centered at S are invertible. □

Lemma 1.3.7 *Let* X *be a compact Hausdorff space and let* n *be a natural number. Suppose that* S *and* T *are in* $\mathrm{GL}(n, C(X))$ *and satisfy the inequality*

$$\|S - T\|_\infty < \frac{1}{2\|S^{-1}\|_\infty}.$$

Then

$$\|S^{-1} - T^{-1}\|_\infty < 2\|S^{-1}\|_\infty^2 \|S - T\|_\infty .$$

Proof From the hypothesized inequality, we have

$$\|I_n - S^{-1}T\|_\infty = \|S^{-1}(S - T)\|_\infty \le \|S^{-1}\|_\infty \|S - T\|_\infty < \frac{1}{2},$$

and Proposition 1.3.4 yields

$$\left\|\mathsf{T}^{-1}\right\|_\infty \le \left\|(\mathsf{S}^{-1}\mathsf{T})^{-1}\right\|_\infty \left\|\mathsf{S}^{-1}\right\|_\infty$$

$$\le \frac{1}{1 - \left\|I_n - \mathsf{S}^{-1}\mathsf{T}\right\|_\infty} \cdot \left\|\mathsf{S}^{-1}\right\|_\infty < 2\left\|\mathsf{S}^{-1}\right\|_\infty .$$

Therefore

$$\left\|\mathsf{S}^{-1} - \mathsf{T}^{-1}\right\|_\infty = \left\|\mathsf{S}^{-1}(\mathsf{T} - \mathsf{S})\mathsf{T}^{-1}\right\|_\infty$$

$$\le \left\|\mathsf{S}^{-1}\right\|_\infty \left\|\mathsf{T}^{-1}\right\|_\infty \left\|\mathsf{T} - \mathsf{S}\right\|_\infty$$

$$< 2\left\|\mathsf{S}^{-1}\right\|_\infty^2 \left\|\mathsf{S} - \mathsf{T}\right\|_\infty .$$

$\square$

Proposition 1.3.8 *For every compact Hausdorff space X and natural number n, the group $\mathrm{GL}(n, C(X))$ is a topological group; i.e., multiplication and inversion are continuous operations.*

Proof The only nontrivial point to check is that the map $\mathsf{S} \mapsto \mathsf{S}^{-1}$ is continuous, and this follows immediately from Lemma 1.3.7. $\square$

The group $\mathrm{GL}(n, C(X))$ has a normal subgroup in which we are particularly interested.

Definition 1.3.9 *Let X be compact Hausdorff and let n be a natural number. We let $\mathrm{GL}(n, C(X))_0$ denote the connected component of I_n in $\mathrm{GL}(n, C(X))$; i.e., the maximal connected subset of $\mathrm{GL}(n, C(X))$ that contains the identity.*

Proposition 1.3.10 *Let X be compact Hausdorff and let n be a natural number. Then $\mathrm{GL}(n, C(X))_0$ is a normal (in the sense of group theory) subgroup of $\mathrm{GL}(n, C(X))$.*

Proof Proposition 1.3.5 implies that $\mathrm{GL}(n, C(X))$ is locally path connected and therefore the connected components and the path components of $\mathrm{GL}(n, C(X))$ coincide. By definition $\mathrm{GL}(n, C(X))_0$ contains I_n. Take S and T in $\mathrm{GL}(n, C(X))_0$, let $\{\mathsf{S}_t\}$ be a continuous path in $\mathrm{GL}(n, C(X))_0$ from $I_n = \mathsf{S}_0$ to $\mathsf{S} = \mathsf{S}_1$, and let $\{\mathsf{T}_t\}$ be a continuous path in $\mathrm{GL}(n, C(X))_0$ from $I_n = \mathsf{T}_0$ to $\mathsf{T} = \mathsf{T}_1$. Then $\{\mathsf{S}_t\mathsf{T}_t\}$ is a continuous path from I_n to $\mathsf{S}\mathsf{T}$ and $\{\mathsf{S}_t^{-1}\}$ is a continuous path from I_n to S^{-1}, whence $\mathrm{GL}(n, C(X))_0$ is a subgroup. Furthermore, for any R in

$GL(n, C(X))$, we have a continuous path $\{RS_tR^{-1}\}$ from I_n to RSR^{-1}, and therefore $GL(n, C(X))_0$ is a normal subgroup of $GL(n, C(X))$. $\square$

Proposition 1.3.11 *For all natural numbers n, the subgroup $GL(n, \mathbb{C})_0$ equals $GL(n, \mathbb{C})$.*

Proof Take S in $GL(n, \mathbb{C})$. Then S is similar to an upper triangular matrix T. For each entry of T that lies above the main diagonal, multiply by $1 - t$ and let t go from 0 to 1; this gives us a homotopy in $GL(n, \mathbb{C})$ from T to a diagonal matrix D. Let $\lambda_1, \lambda_2, \ldots, \lambda_n$ be the diagonal entries of D that are not equal to 1; because D is invertible, none of these diagonal entries is zero. For each $1 \leq k \leq n$, choose a homotopy $\{\lambda_{k,t}\}$ from λ_k to 1; these homotopies determine a homotopy from D to I_n. Thus D and T are in $GL(n, \mathbb{C})_0$, and because $GL(n, \mathbb{C})_0$ is a normal subgroup of $GL(n, \mathbb{C})$, the matrix S is in $GL(n, \mathbb{C})_0$ as well. Our choice of S was arbitrary and thus $GL(n, \mathbb{C})_0 = GL(n, \mathbb{C})$. $\square$

To further understand the structure of $GL(n, C(X))_0$, we need an alternate description of it.

Proposition 1.3.12 *Let X be compact Hausdorff, let n be a natural number, and suppose B is an element of $M(n, C(X))$. Then the exponential*

$$\exp B = \sum_{k=0}^{\infty} \frac{B^k}{k!}$$

of B is well defined.

Proof For all natural numbers N, we have the inequality

$$\left\| \sum_{k=N}^{\infty} \frac{B^k}{k!} \right\|_{\infty} \leq \sum_{k=N}^{\infty} \frac{\|B\|_{\infty}^k}{k!}.$$

The desired result follows from Proposition 1.2.4 and the fact that the Maclaurin series for e^x has an infinite radius of convergence. $\square$

Proposition 1.3.13 *Let X be a compact Hausdorff space and let n be a natural number. Then the exponential map is a continuous function from $M(n, C(X))$ to $GL(n, C(X))$.*

Proof A straightforward computation shows that $\exp B \exp(-B) = I_n$ and so $\exp(-B) = (\exp B)^{-1}$. Therefore the range of exp is contained in $\mathrm{GL}(n, C(X))$. □

In general, $\exp : \mathrm{M}(n, C(X)) \longrightarrow \mathrm{GL}(n, C(X))$ is not a group homomorphism; if A and B do not commute, $\exp(A + B)$ is not necessarily equal to $\exp A \exp B$.

Lemma 1.3.14 *Let X be a compact Hausdorff space and let n be a natural number. If S is in $\mathrm{M}(n, C(X))$ and $\|I_n - S\|_\infty < 1$, then $S = \exp B$ for some B in $\mathrm{M}(n, C(X))$.*

Proof The Taylor series expansion of $\log x$ centered at 1 is

$$-\sum_{k=1}^{\infty} \frac{(1-x)^k}{k}$$

and has radius of convergence 1. Set

$$B = -\sum_{k=1}^{\infty} \frac{(I_n - S)^k}{k}.$$

Compute $\exp B$ and simplify to obtain S. □

Theorem 1.3.15 *Let X be a compact Hausdorff space and let n be a natural number. Then the group $\mathrm{GL}(n, C(X))_0$ is precisely the set of finite products of elements of the form $\exp B$ for B in $\mathrm{M}(n, C(X))$.*

Proof Let $\mathcal{F}$ denote the set described in the statement of the theorem and take

$$S = \exp B_1 \exp B_2 \cdots \exp B_m$$

in $\mathcal{F}$. Then

$$\exp(-B_m) \exp(-B_{m-1}) \cdots \exp(-B_1)$$

is an inverse to S, whence S is in $\mathrm{GL}(n, C(X))$. The product

$$S_t = \exp(tB_1) \exp(tB_2) \cdots \exp(tB_m)$$

defines a homotopy in $\mathrm{GL}(n, C(X))$ from I_n to S and thus S is in $\mathrm{GL}(n, C(X))_0$.

Observe that $\mathcal{F}$ is a subgroup of $\mathrm{GL}(n, C(X))$. Lemma 1.3.14 implies that there is an open neighborhood U of I_n that lies entirely in $\mathcal{F}$. Group multiplication by any element of $\mathcal{F}$ is a homeomorphism from $\mathcal{F}$ to itself, so $\mathcal{F}$ is a union of open sets and therefore open. Each left coset of $\mathcal{F}$ in $\mathrm{GL}(n, C(X))$ is homeomorphic to $\mathcal{F}$ and is therefore also open. The left cosets partition $\mathrm{GL}(n, C(X))$ into disjoint sets, so each left coset, and in particular $\mathcal{F}$, is both open and closed. Therefore $\mathcal{F}$ is a connected component of $\mathrm{GL}(n, C(X))$, whence $\mathcal{F} = \mathrm{GL}(n, C(X))_0$. $\qquad \square$

Proposition 1.3.16 *Let n be a natural number and suppose that A is a nonempty closed subspace of a compact Hausdorff space X. Then the restriction of elements of $\mathrm{M}(n, C(X))$ to $\mathrm{M}(n, C(A))$ determines a continuous surjective group homomorphism from $\mathrm{GL}(n, C(X))_0$ to $\mathrm{GL}(n, C(A))_0$.*

Proof Let ρ denote the restriction map from $\mathrm{M}(n, C(X))$ to $\mathrm{M}(n, C(A))$. Then ρ clearly maps $\mathrm{GL}(n, C(X))_0$ into $\mathrm{GL}(n, C(A))_0$. To show that ρ maps $\mathrm{GL}(n, C(X))_0$ onto $\mathrm{GL}(n, C(A))_0$, we first show that ρ maps $\mathrm{M}(n, C(X))$ onto $\mathrm{M}(n, C(A))$. Choose an element B of $\mathrm{M}(n, C(A))$. By applying the Tietze extension theorem to each matrix entry of B we construct a matrix B' in $\mathrm{M}(n, C(X))$ such that $\rho(\mathsf{B}') = \mathsf{B}$.

Now take S in $\mathrm{GL}(n, C(A))_0$. By Theorem 1.3.15 we can write

$$\mathsf{S} = \exp \mathsf{B}_1 \exp \mathsf{B}_2 \cdots \exp \mathsf{B}_m$$

for some B_1, B_2, ..., B_m in $\mathrm{M}(n, C(A))$. For each $1 \le k \le m$, choose an element B'_k in $\mathrm{M}(n, C(X))$ with the property that $\rho(\mathsf{B}'_k) = \mathsf{B}_k$. Then

$$\begin{aligned}
\rho\left(\exp \mathsf{B}'_1 \exp \mathsf{B}'_2 \cdots \exp \mathsf{B}'_m\right) &= \exp \rho(\mathsf{B}'_1) \exp \rho(\mathsf{B}'_2) \cdots \exp \rho(\mathsf{B}'_m) \\
&= \exp \mathsf{B}_1 \exp \mathsf{B}_2 \cdots \exp \mathsf{B}_m \\
&= \mathsf{S}.
\end{aligned}$$

$\qquad \square$

There is a special case of Proposition 1.3.16 that we will use repeatedly in Chapter 2.

Corollary 1.3.17 *Suppose that A is a nonempty closed subspace of a compact Hausdorff space X and let n be a natural number. Then for each S in $\mathrm{GL}(n, C(A))$, there exists T in $\mathrm{GL}(2n, C(X))_0$ such that $\mathsf{T}|A = \mathrm{diag}(\mathsf{S}, \mathsf{S}^{-1})$.*

Proof Proposition 1.3.16 implies that we need only find a homotopy in $\mathrm{GL}(2n, C(A))$ from $I_{2n} = \mathrm{diag}(\mathsf{S}\mathsf{S}^{-1}, I_n)$ to $\mathrm{diag}(\mathsf{S}, \mathsf{S}^{-1})$; Proposition 1.3.3 provides such a homotopy. $\qquad \square$

We will also need the following result in the next chapter.

Proposition 1.3.18 *Let X be compact Hausdorff and let n be a natural number. A matrix B in $\mathrm{M}(2n, C(X))$ commutes with $\mathrm{diag}(I_n, 0_n)$ if and only if $\mathsf{B} = \mathrm{diag}(\mathsf{B}_{11}, \mathsf{B}_{22})$ for some matrices B_{11} and B_{22} in $\mathrm{M}(n, C(X))$. Moreover, if B commutes with the matrix $\mathrm{diag}(I_n, 0_n)$, then B is invertible if and only if B_{11} and B_{22} are.*

Proof Write B in the form

$$\mathsf{B} = \begin{pmatrix} \mathsf{B}_{11} & \mathsf{B}_{12} \\ \mathsf{B}_{21} & \mathsf{B}_{22} \end{pmatrix}.$$

Then

$$\mathsf{B}\,\mathrm{diag}(I_n, 0_n) = \begin{pmatrix} \mathsf{B}_{11} & \mathsf{B}_{12} \\ \mathsf{B}_{21} & \mathsf{B}_{22} \end{pmatrix} \begin{pmatrix} I_n & 0 \\ 0 & 0_n \end{pmatrix} = \begin{pmatrix} \mathsf{B}_{11} & 0 \\ \mathsf{B}_{21} & 0 \end{pmatrix}$$

and

$$\mathrm{diag}(I_n, 0_n)\mathsf{B} = \begin{pmatrix} I_n & 0 \\ 0 & 0_n \end{pmatrix} \begin{pmatrix} \mathsf{B}_{11} & \mathsf{B}_{12} \\ \mathsf{B}_{21} & \mathsf{B}_{22} \end{pmatrix} = \begin{pmatrix} \mathsf{B}_{11} & \mathsf{B}_{12} \\ 0 & 0 \end{pmatrix};$$

the result then follows easily. $\qquad\qquad\square$

1.4 Idempotents

In this section we define one of the primary algebraic objects of study in K-theory and look at some of its properties.

Definition 1.4.1 *Let X be a compact Hausdorff space and let n be a natural number. An* idempotent *over X is an element E of $\mathrm{M}(n, C(X))$ with the property that $\mathsf{E}^2 = \mathsf{E}$.*

We can define idempotents for any ring, but Definition 1.4.1 will serve our purposes.

Example 1.4.2 *For any compact Hausdorff space X, a matrix that consists of 1s and 0s on the main diagonal and is 0 elsewhere is an idempotent.*

Example 1.4.3 *Let S^2 be the unit sphere in $\mathbb{R}^3$ centered at the origin. Then the matrix*

$$\frac{1}{2}\begin{pmatrix} 1+x & y+iz \\ y-iz & 1-x \end{pmatrix}$$

is an idempotent over S^2.

Example 1.4.4 *For each natural number m, consider*

$$S^m = \{(x_1, x_2, \ldots, x_{m+1}) \in \mathbb{R}^{m+1} : x_1^2 + x_2^2 + \cdots + x_{m+1}^2 = 1\},$$

the m-dimensional sphere of radius 1 centered at the origin. Then

$$\begin{pmatrix} x_1^2 & x_1 x_2 & x_1 x_3 & \cdots & x_1 x_{m+1} \\ x_2 x_1 & x_2^2 & x_2 x_3 & \cdots & x_2 x_{m+1} \\ \vdots & \vdots & \vdots & \ddots & \vdots \\ x_{m+1} x_1 & x_{m+1} x_2 & x_{m+1} x_3 & \cdots & x_{m+1}^2 \end{pmatrix}$$

is an idempotent over S^m.

Example 1.4.5 *Consider S^1 as the unit circle in $\mathbb{C}$ and view the two-torus $\mathbb{T}^2$ as $S^1 \times [0, 1]$ with the usual identification of the ends: $(z, 0) \sim (z, 1)$ for each z in S^1. Set*

$$\mathsf{E}(z, t) = \mathsf{Rot}(t) \operatorname{diag}(z, 1) \mathsf{Rot}^{-1}(t).$$

Now, $\mathsf{E}(z, 0) = \operatorname{diag}(z, 1) \neq \operatorname{diag}(1, z) = \mathsf{E}(z, 1)$, so E is not an element of $\mathrm{M}(2, C(\mathbb{T}^2))$. However, the matrix $\mathsf{E}(z, t) \operatorname{diag}(1, 0) \mathsf{E}(z, t)^{-1}$ is a well defined idempotent over $\mathbb{T}^2$.

Definition 1.4.6 *Let X be compact Hausdorff and let n be a natural number. Idempotents E_0 and E_1 in $\mathrm{M}(n, C(X))$ are homotopic if there exists a homotopy $\{\mathsf{E}_t\}$ of idempotents in $\mathrm{M}(n, C(X))$ from E_0 to E_1. If E_0 and E_1 are homotopic, we write $\mathsf{E}_0 \sim_h \mathsf{E}_1$.*

Definition 1.4.7 *Let X be compact Hausdorff and let n be a natural number. Idempotents E_0 and F in $\mathrm{M}(n, C(X))$ are similar if $\mathsf{F} = \mathsf{SES}^{-1}$ for some S in $\mathrm{GL}(n, C(X))$. If E and F are similar, we write $\mathsf{F} \sim_s \mathsf{E}$.*

Similarity and homotopy are both equivalence relations; our next task is to see how similarity and homotopy are related.

Lemma 1.4.8 *Let X be compact Hausdorff and let n be a natural number. Take idempotents E and F in $\mathrm{M}(n, C(X))$, and suppose that*

$$\|\mathsf{F} - \mathsf{E}\|_\infty < \frac{1}{\|2\mathsf{F} - I_n\|_\infty}.$$

Then $\mathsf{S} = I_n - \mathsf{E} - \mathsf{F} + 2\mathsf{EF}$ is invertible and $\mathsf{E} = \mathsf{SFS}^{-1}$. In particular, $\mathsf{F} \sim_s \mathsf{E}$.

Proof We have

$$\|I_n - \mathsf{S}\|_\infty = \|\mathsf{E} + \mathsf{F} - 2\mathsf{EF}\|_\infty = \|(\mathsf{F} - \mathsf{E})(2\mathsf{F} - I_n)\|_\infty$$
$$\leq \|\mathsf{F} - \mathsf{E}\|_\infty \, \|2\mathsf{F} - I_n\|_\infty < 1.$$

By Proposition 1.3.4, the matrix S is invertible. The products SF and ES are both equal to EF, so $\mathsf{ES} = \mathsf{SF}$ and $\mathsf{E} = \mathsf{SFS}^{-1}$. □

Proposition 1.4.9 *Let X be a compact Hausdorff space, let n be a natural number, and suppose E_0 and E_1 are homotopic idempotents in $\mathrm{M}(n, C(X))$. Then E_0 and E_1 are similar.*

Proof Choose a homotopy of idempotents from E_0 to E_1; for clarity of notation in what follows, we will write the homotopy as $\{\mathsf{E}(t)\}$ instead of $\{\mathsf{E}_t\}$. Set

$$R = \sup\{\|2\mathsf{E}(t) - I_n\|_\infty : 0 \leq t \leq 1\},$$

and choose points $0 = t_0 < t_1 < t_2 < \cdots < t_K = 1$ so that

$$\|\mathsf{E}(t_{k-1}) - \mathsf{E}(t_k)\|_\infty < \frac{1}{R}$$

for all $1 \leq k \leq K$. By Lemma 1.4.8, we know that $\mathsf{E}(t_{k-1}) \sim_s \mathsf{E}(t_k)$ for all k and therefore $\mathsf{E}_0 \sim_s \mathsf{E}_1$. □

Proposition 1.4.10 *Let X be compact Hausdorff and let n be a natural number. If E and F in $\mathrm{M}(n, C(X))$ are similar, then $\mathrm{diag}(\mathsf{E}, 0_n)$ and $\mathrm{diag}(\mathsf{F}, 0_n)$ are homotopic idempotents in $\mathrm{M}(2n, C(X))$.*

Proof Choose S in $\mathrm{GL}(n, C(X))$ so that $\mathsf{F} = \mathsf{SES}^{-1}$. For $0 \leq t \leq 1$, define

$$\mathsf{T}_t = \mathrm{diag}(\mathsf{S}, I_n)\mathsf{Rot}(t)\,\mathrm{diag}(\mathsf{S}^{-1}, I_n)\mathsf{Rot}^{-1}(t).$$

Then $\{\mathsf{T}_t\,\mathrm{diag}(\mathsf{E}, 0_n)\mathsf{T}_t^{-1}\}$ is the desired homotopy of idempotents. □

Proposition 1.4.10 suggests that if we want homotopy classes and similarity classes to coincide, we should consider matrices of all sizes. The next definition makes this idea precise.

Definition 1.4.11 *For every compact Hausdorff space X, define an equivalence relation $\sim$ on $\bigcup_{n \in \mathbb{N}} \mathrm{M}(n, C(X))$ by declaring that*

$$\mathsf{B} \sim \mathrm{diag}(\mathsf{B}, 0)$$

for all natural numbers n and matrices B in $\mathrm{M}(n, C(X))$. The set of equivalence classes of $\sim$ is denoted $\mathrm{M}(C(X))$.

The construction in Definition 1.4.11 is an example of a *direct limit*; we often indicate this by writing

$$\mathrm{M}(C(X)) = \varinjlim \mathrm{M}(n, C(X)).$$

We will usually identify $\mathrm{M}(n, C(X))$ with its image in $\mathrm{M}(C(X))$. We can view an element of $\mathrm{M}(C(X))$ as a countably infinite matrix with entries in $C(X)$ and all but finitely many entries equal to 0.

Proposition 1.4.12 *Let X be compact Hausdorff. Then $\mathrm{M}(C(X))$ is an algebra without unit under the operations of matrix addition and multiplication and scalar multiplication.*

Proof Let A and B be elements of $\mathrm{M}(C(X))$, and choose a natural number n large enough so that A and B can be viewed as elements of $\mathrm{M}(n, C(X))$. Then we can define $\mathsf{A} + \mathsf{B}$ and $\mathsf{A}\mathsf{B}$ as elements of $\mathrm{M}(n, C(X))$ in the usual way, and these operations are compatible with the equivalence relation that defines $\mathrm{M}(C(X))$. Similarly, for any complex number λ, we can consider $\lambda\mathsf{A}$ as an element of $\mathrm{M}(C(X))$. Finally, if $\mathrm{M}(C(X))$ had a multiplicative identity, then it would have to be the countably infinite matrix with 1s on the main diagonal and 0s everywhere else; this matrix is not an element of $\mathrm{M}(C(X))$. $\square$

Definition 1.4.13 *Given a compact Hausdorff space X, the* direct limit topology *on $\mathrm{M}(C(X))$ is the topology whose basis consists of images of open sets in $\mathrm{M}(n, C(X))$ for every natural number n.*

If E and F are homotopic idempotents in $\mathrm{M}(n, C(X))$, then $\mathrm{diag}(\mathsf{E}, 0)$ and $\mathrm{diag}(\mathsf{F}, 0)$ are homotopic idempotents in $\mathrm{M}(n + 1, C(X))$. Thus the notion of homotopic idempotents in $\mathrm{M}(C(X))$ is well defined and we can make the following definition.

Definition 1.4.14 *Let X be compact Hausdorff. The collection of similarity classes (i.e., the equivalence classes of $\sim_s$) of idempotents in $\mathrm{M}(C(X))$ is denoted $\mathrm{Idem}(C(X))$. Given an idempotent E, we denote its similarity class by $[\mathsf{E}]$.*

Propositions 1.4.9 and 1.4.10 imply that we can alternately define $\mathrm{Idem}(C(X))$ to be the set of homotopy classes of idempotents in the algebra $\mathrm{M}(C(X))$.

Is it possible that each of the idempotents in Examples 1.4.2 through 1.4.5 is similar to I_n for some n? Later, we will be able to show that the answer to this question is no. However, at this stage we have no tools for determining when two idempotents determine distinct elements of $\mathrm{Idem}(C(X))$. This is an important topic that we take up in Chapter 4.

1.5 Vector bundles

In this section we look at families of vector spaces over a topological space. We require that the vector spaces vary continuously and that they be locally trivial in a sense that will be described shortly. We start out working over arbitrary topological spaces, but to obtain a well behaved theory, we eventually restrict our attention to compact Hausdorff spaces.

Definition 1.5.1 *Let X be a topological space. A family of vector spaces over X is a topological space V and a continuous surjective map $\pi : V \longrightarrow X$, called the projection, such that for each x in X the inverse image $\pi^{-1}(x)$ of x is a finite-dimensional complex vector space whose addition and scalar multiplication are continuous in the subspace topology on $\pi^{-1}(x)$.*

A family of vector spaces is sometimes written (V, π, X), but we will often write (V, π) or just V if there is no possibility of confusion. We usually will write V_x for $\pi^{-1}(x)$; this vector space is called the *fiber* of V over $x \in X$. Occasionally, we will write V as $\{V_x\}_{x \in X}$. The topological space X is called the *base* of V.

Do not confuse the projection of a family of vector spaces onto its base with the notion of an orthogonal projection of a vector space onto a vector subspace.

Example 1.5.2 *For any natural number n, the space $X \times \mathbb{C}^n$ is a family of vector spaces over X; the projection $\pi : X \times \mathbb{C}^n \longrightarrow X$ is $\pi(x, \vec{z}) = x$. We denote this family $\Theta^n(X)$.*

Example 1.5.3 *For any topological space X and natural number n, we can produce a multitude of families of vector spaces by choosing for each $x \in X$ any vector subspace V_x whatsoever of $\Theta^n(X)_x$.*

Example 1.5.3 shows that the vector spaces that make up a family of vector spaces can vary wildly and therefore it is difficult to prove very much at this level of generality. Fortunately, many naturally occurring examples of families of vector spaces have additional structure, which we describe below. We begin with some definitions.

Definition 1.5.4 *Let V and W be families of vector spaces over a topological space X. A continuous function $\gamma : V \longrightarrow W$ is a* homomorphism *of families if for each $x \in X$ the map γ restricts to a vector space homomorphism $\gamma_x : V_x \longrightarrow W_x$. If γ is a homeomorphism (so that, in particular, γ_x is a vector space isomorphism for each $x \in X$), we call γ an* isomorphism *of families.*

Definition 1.5.5 *A family of vector spaces V over a topological space X is* trivial *if V is isomorphic to $\Theta^n(X)$ for some natural number n.*

The next definition is a special case of a construction we shall consider later in this section.

Definition 1.5.6 *Let (V, π, X) be a family of vector spaces over a topological space X and let A be a subspace of X. The family $(\pi^{-1}(A), \pi, A)$ of vector spaces over A is called the* restriction *of V to A and is denoted $V|A$.*

Definition 1.5.7 *A* vector bundle *over a topological space X is a family of vector spaces V with the property that for each x in X, there is an open neighborhood U of x in X such that the restriction $V|U$ of V to U is trivial.*

When V is a vector bundle, the dimension of V_x is a locally constant function of x. When the dimension of all the fibers V_x is actually constant (in particular, when X is connected), we call this number the *rank* of the vector bundle V. If V is a rank one bundle, we say that V is a *(complex) line bundle.*

A homomorphism between vector bundles is often called a *bundle homomorphism*, and similarly for isomorphisms.

Definition 1.5.8 *Let X be a topological space. The collection of isomorphism classes of vector bundles over X is denoted $\mathrm{Vect}(X)$. If V is a vector bundle over X, we denote its isomorphism class $[V]$.*

We now look at some examples of vector bundles.

Example 1.5.9 *For each natural number n, the family of vector spaces $\Theta^n(X)$ is a rank n vector bundle over X.*

Example 1.5.10 *Define an equivalence relation $\sim_p$ on $\mathbb{C}^2$ by setting $(z_1, z_2) \sim_p (\lambda z_1, \lambda z_2)$ for every nonzero complex number λ. If we give $\mathbb{C}^2$ its usual topology and endow the set of the equivalence classes of $\sim_p$ with the quotient topology, we obtain a topological space $\mathbb{CP}^1$ called* complex projective space. *The equivalence class of a point (z_1, z_2) is denoted $[z_1, z_2]$, and these equivalence classes are collectively known as* homogeneous coordinates *for $\mathbb{CP}^1$. Every point of $\mathbb{CP}^1$ is a one-dimensional complex vector subspace of $\mathbb{C}^2$, and the disjoint union of these vector subspaces is a family H^* of vector subspaces of $\Theta^2(\mathbb{CP}^1)$ called the* tautological line bundle over $\mathbb{CP}^1$.

To show that H^* is a line bundle, set

$$U_1 = \{[z_1, z_2] \in \mathbb{C}P^1 : z_1 \neq 0\}$$
$$U_2 = \{[z_1, z_2] \in \mathbb{C}P^1 : z_2 \neq 0\}.$$

Then $\mathbb{CP}^1$ is the union of U_1 and U_2. The map $\phi_1 : H^*|U_1 \longrightarrow \Theta^1(U_1)$ defined by $\phi_1[z_1, z_2] = \big([z_1, z_2], z_2/z_1\big)$ is a bundle isomorphism. Similarly, the map $\phi_2 : H^*|U_2 \longrightarrow \Theta^1(U_2)$ given by the formula $\phi_2[z_1, z_2] = \big([z_1, z_2], z_1/z_2\big)$ is a bundle isomorphism.

More generally, for any natural number n we can define the equivalence relation

$$(z_1, z_2, \ldots, z_{n+1}) \sim_p (\lambda z_1, \lambda z_2, \ldots, \lambda z_{n+1}), \qquad \lambda \neq 0$$

on $\mathbb{C}^{n+1}$ and take equivalence classes to get complex projective n-space $\mathbb{CP}^n$. *As in the case of $\mathbb{CP}^1$, we have a tautological line bundle over $\mathbb{CP}^n$.*

The unusual name for the vector bundle in Example 1.5.10 comes from algebraic geometry; we will consider a generalization of this construction in Section 3.6.

To obtain other examples of vector bundles, it is helpful to be able to create new vector bundles out of old ones. There are a variety of such constructions; roughly speaking, any construction that we can do with vector spaces has an analogue for vector bundles. We consider a few of these constructions below.

Construction 1.5.11 *Let (V, π) be a vector bundle over a topological space X and let W be a topological subspace of V. If $(W, \pi|W)$ is a vector bundle over X, then we call W a* subbundle *of V.*

Construction 1.5.12 *Let X and Y be topological spaces, let (V, π) be a vector bundle over Y, and suppose $\phi : X \longrightarrow Y$ is a continuous map. Define*

$$\phi^* V = \{(x, v) \in X \times V : \phi(x) = \pi(v)\}.$$

Endow $\phi^* V$ with the subspace topology it inherits from $X \times V$. Then $\phi^* V$ is a family of vector spaces over X called the pullback *of V by ϕ; the projection $\phi^* \pi : \phi^* V \longrightarrow X$ is simply $(\phi^* \pi)(x, v) = x$. The family $\phi^* V$ is locally trivial, for if $V|U$ is trivial over an open subset U of Y, then $\phi^* V|\phi^{-1}(U)$ is trivial as well. Thus $\phi^* V$ is a vector bundle over X.*

The restriction of a vector bundle to a subspace is a special case of the pullback construction.

Construction 1.5.13 *Let X_1 and X_2 be topological spaces and let (V_1, π_1) and (V_2, π_2) be vector bundles over X_1 and X_2 respectively. Let $V_1 \times V_2$ be the product of V_1 and V_2 as topological spaces and define $\pi_\boxplus : V_1 \times V_2 \longrightarrow X_1 \times X_2$ by the formula $\pi_\boxplus(v_1, v_2) = (\pi_1(v_1), \pi_2(v_2))$. Then $(V_1 \times V_2, \pi_\boxplus)$ is a vector bundle over $X_1 \times X_2$ called the* external Whitney sum *of V_1 and V_2; we write this bundle $V_1 \boxplus V_2$. More generally, given a finite number $V_1, V_2, \ldots, V_m$ of vector bundles over topological spaces $X_1, X_2, \ldots, X_m$ respectively, we can form the external Whitney sum $V_1 \boxplus V_2 \boxplus \cdots \boxplus V_m$ over $X_1 \times X_2 \times \cdots \times X_m$.*

Construction 1.5.14 *Let (V_1, π_1) and (V_2, π_2) be vector bundles over the same base space X, and consider the subspace*

$$V_1 \oplus V_2 = \{(v_1, v_2) \in V_1 \times V_2 : \pi_1(v_1) = \pi_2(v_2)\}$$

of $V_1 \times V_2$. If we define $\pi_\oplus : V_1 \oplus V_2 \longrightarrow X$ as $(\pi_\oplus)(v_1, v_2) = \pi_1(v_1)$, then we obtain a vector bundle $(V_1 \oplus V_2, \pi_\oplus)$ over X called the internal Whitney sum *of V_1 and V_2. An alternate way of defining internal Whitney sum is this: let $\Delta : X \longrightarrow X \times X$ be the* diagonal map; *i.e., $\Delta(x) = (x, x)$ for every x in X. Then $V_1 \oplus V_2 = \Delta^*(V_1 \boxplus V_2)$.*

As in the case of the external Whitney sum, we can take the internal Whitney sum $V_1 \oplus V_2 \oplus \cdots \oplus V_m$ of any finite number of vector bundles over X.

Construction 1.5.15 *Let V be a vector bundle over a topological space X. The* dual vector bundle *of V is the set V^* of vector bundle homomorphisms from V to $\Theta^1(X)$. The dual vector bundle of V is a family of vector spaces over X because for each x in X, the elements of V^* restrict to linear maps from V_x to $\mathbb{C}$. To see that V^* is locally trivial, let U be a connected open subset of X such that $V|U$ is a trivial vector bundle of rank n. Then $V^*|U$ is isomorphic to the collection of bundle homomorphisms from $\Theta^n(U)$ to $\Theta^1(U)$, which in turn is isomorphic to $\Theta^n(U)$.*

If we take the definition of vector bundle and replace the complex numbers $\mathbb{C}$ with the real numbers $\mathbb{R}$, we obtain a *real vector bundle*. The prototypical example of a real vector bundle is the tangent bundle of a smooth manifold. We can manufacture a complex vector bundle out of a real one by *complexification*.

Construction 1.5.16 *Let V be a real vector bundle over a topological space X and define a real vector bundle $V \oplus V$ over X by (the real analogue of) internal Whitney sum. Each fiber of $V \oplus V$ is a complex vector space via the scalar multiplication*

$$(a + bi)(v_1, v_2) = (av_1 - bv_2, bv_1 + av_2),$$

which makes $V \oplus V$ into a complex vector bundle denoted $V \otimes \mathbb{C}$. Elements of $(V \otimes \mathbb{C})_x$ are usually written as formal sums $v_1 + iv_2$, where v_1, v_2 are in V_x.

At the end of this section we consider one more vector bundle construction. First, we must establish some preliminary results.

Lemma 1.5.17 *Let m and n be natural numbers, suppose that X is a Hausdorff space, not necessarily compact, and let $\mathrm{Map}(X, \mathrm{M}(m, n, \mathbb{C}))$ denote the set of continuous functions from X to $\mathrm{M}(m, n, \mathbb{C})$. For every f in $\mathrm{Map}(X, \mathrm{M}(m, n, \mathbb{C}))$, define a bundle homomorphism*

$$\Gamma(f) : \Theta^n(X) \longrightarrow \Theta^m(X)$$

by the formula

$$\Gamma(f)(x, \vec{z}) = f(x)\vec{z}.$$

Let $\mathrm{Hom}(\Theta^n(X), \Theta^m(X))$ *be the set of bundle homomorphisms from* $\Theta^n(X)$ *to* $\Theta^m(X)$. *Then* Γ *is a bijection from* $\mathrm{Map}(X, \mathrm{M}(m, n, \mathbb{C}))$ *to* $\mathrm{Hom}(\Theta^n(X), \Theta^m(X))$.

Proof The only nonobvious point to establish is surjectivity. Equip $\mathbb{C}^m$ and $\mathbb{C}^n$ with their standard vector space bases and standard inner products and define $p : \Theta^n(X) \longrightarrow \mathbb{C}^n$ as $p(x, \vec{z}) = \vec{z}$. Given a bundle homomorphism $\delta : \Theta^n(X) \longrightarrow \Theta^m(X)$, define $f : X \longrightarrow \mathrm{M}(m, n, \mathbb{C})$ so that the (i, j) entry is

$$f_{ij}(x) = \langle p\left(\delta(x, e_j)\right), e_i \rangle$$

for all x in X. Then f is in $\mathrm{Map}(X, \mathrm{M}(m, n, \mathbb{C}))$, and $\Gamma(f) = \delta$. $\square$

We could identify $\mathrm{Map}(X, \mathrm{M}(m, n, \mathbb{C}))$ with $\mathrm{M}(m, n, C(X))$. However, we will find it helpful to preserve this distinction in the proofs of several results to follow.

Definition 1.5.18 *Let* $\mathcal{U} = \{U_1, U_2, \ldots, U_l\}$ *be a finite open cover of a topological space* X. *A partition of unity subordinate to* $\mathcal{U}$ *is a collection of continuous functions* $p_k : X \longrightarrow [0, 1]$ *for* $1 \le k \le l$ *such that:*

(i) $p_k(x) = 0$ *for* $x \in X \backslash U_k$;
(ii) $\sum_{k=1}^{l} p_k(x) = 1$ *for all* x *in* X.

We could define a partition of unity subordinate to an infinite open cover, but Definition 1.5.18 will suffice for our purposes. Any normal topological space admits partitions of unity, so in particular we can always find a partition of unity for any finite open cover of a compact Hausdorff space.

Proposition 1.5.19 *Let* A *be a closed subspace of a compact Hausdorff space* X, *let* V *and* W *be vector bundles over* X, *and suppose* $\sigma : V|A \longrightarrow W|A$ *is a bundle homomorphism. Then* σ *can be extended to a bundle homomorphism* $\tilde{\sigma} : V \longrightarrow W$.

Proof Without loss of generality, we assume that X is connected; otherwise, work with X one component at a time. Suppose the rank of V is n and the rank of W is m. Choose a finite open cover $\mathcal{U} = \{U_1, U_2, \ldots, U_l\}$ of X with the property that for each $1 \le k \le l$, the vector bundles $V|\overline{U}_k$ and $W|\overline{U}_k$ are isomorphic to $\Theta^n(\overline{U}_k)$ and $\Theta^m(\overline{U}_k)$ respectively. For each k, let σ_k denote the map from $V|(A \cap \overline{U}_k)$ to $W|(A \cap \overline{U}_k)$ defined by restricting the domain and range of σ. If $A \cap \overline{U}_k$ is empty, define $\tilde{\sigma}_k : V \longrightarrow W$ to be the zero homomorphism. Otherwise, identify $V|(A \cap \overline{U}_k)$ and $W|(A \cap \overline{U}_k)$ with $\Theta^n(A \cap \overline{U}_k)$ and $\Theta^m(A \cap \overline{U}_k)$ respectively and apply Lemma 1.5.17 to obtain a map

$$f_k : A \cap \overline{U}_k \longrightarrow \mathrm{M}(m, n, \mathbb{C}).$$

For each $1 \leq i \leq m$ and $1 \leq j \leq n$, the (i, j) entry $(f_k)_{ij}$ of f_k is a continuous function from $A \cap \overline{U}_k$ to $\mathbb{C}$. By the Tietze extension theorem, each $(f_k)_{ij}$ extends to a continuous function $(\hat{f}_k)_{ij} : \overline{U}_k \longrightarrow \mathbb{C}$. The functions $(\hat{f}_k)_{ij}$ collectively define a function

$$\hat{f}_k : \overline{U}_k \longrightarrow \mathrm{M}(m, n, \mathbb{C}),$$

which in turn determines a bundle homomorphism

$$\Gamma(\hat{f}_k) : V_k|\overline{U}_k \cong \Theta^n(\overline{U}_k) \longrightarrow W_k|\overline{U}_k \cong \Theta^m(\overline{U}_k).$$

Let π_V denote the projection from V to X, let $\{p_1, p_2, \ldots, p_l\}$ be a partition of unity subordinate to $\mathcal{U}$, and define $\tilde{\sigma}_k : V \longrightarrow W$ by

$$\tilde{\sigma}_k(v) = \begin{cases} p_k(\pi_V(v))\Gamma(\hat{f}_k)(v) & \text{if } v \in \pi_V^{-1}(U_k) \\ 0 & \text{otherwise.} \end{cases}$$

Then $\tilde{\sigma} = \sum_{k=1}^l \tilde{\sigma}_k$ is a bundle homomorphism that extends σ. $\qquad\square$

Proposition 1.5.20 *Let X be a compact Hausdorff space, let V and W be vector bundles over X, and let $\gamma : V \longrightarrow W$ be a bundle homomorphism. Then the set*

$$\mathcal{O} = \{x \in X : \gamma_x \text{ is an isomorphism}\}$$

is open in X.

Proof We may assume that X is connected. If $\mathcal{O}$ is empty, then it is open, so suppose it is not empty. Then V and W must have the same rank n. For each point x in X, choose a neighborhood U_x of x with the property that $V|U_x$ and $W|U_x$ are both isomorphic to $\Theta^n(U_x)$. Let γ_x be the restriction of γ to U_x and apply Lemma 1.5.17 to get a continuous map f_x from U_x to $\mathrm{M}(n, \mathbb{C})$. Set $\mathcal{O}_x = f_x^{-1}(\mathrm{GL}(n, \mathbb{C}))$. From Corollary 1.3.6 we know the set $\mathrm{GL}(n, \mathbb{C})$ is open in $\mathrm{M}(n, \mathbb{C})$, so $\mathcal{O}_x$ is open in U_x. Moreover, because U_x is open in X, the set $\mathcal{O}_x$ is open in X and the union of the sets $\mathcal{O}_x$ is an open subset of X.

To complete the proof, note that for every x in X and y in U_x the matrix $f_x(y)$ is invertible if and only if γ_y is an isomorphism. Thus

$$\mathcal{O} = \bigcup_{x \in X} \mathcal{O}_x,$$

and therefore $\mathcal{O}$ is open in X.   □

Construction 1.5.21 *Let X be a compact Hausdorff space and suppose A_1 and A_2 are closed subspaces of X whose union is X and whose intersection Z is nonempty. Let V_1 and V_2 be vector bundles over A_1 and A_2 respectively and suppose that we have a bundle isomorphism $\gamma : V_1|Z \longrightarrow V_2|Z$; we call this isomorphism a clutching map. Identify each point v_1 in $V_1|Z$ with its image $\gamma(v_1)$ in $V_2|Z$ and give the resulting set $V_1 \cup_\gamma V_2$ the quotient topology. Then $V_1 \cup_\gamma V_2$ is a family of vector spaces over X.*

In the following proposition, we maintain the notation of Construction 1.5.21.

Proposition 1.5.22 *The family of vector spaces $V_1 \cup_\gamma V_2$ is a vector bundle over X.*

Proof We must show that $V_1 \cup_\gamma V_2$ is locally trivial. Take x in X. If x is in A_1 but not in A_2, then there exists an open neighborhood U of x that does not meet Z. Furthermore, by shrinking U if necessary, we may assume that V_1 is trivial when restricted to U. Then $(V_1 \cup_\gamma V_2)|U \cong V_1|U$ is trivial as well. Similarly, if x is in A_2 but not in A_1, then $V_1 \cup_\gamma V_2$ is trivial over some open neighborhood of x that does not meet A_1. Now suppose x is in Z. The topological space A_2 is normal and thus we can choose an open neighborhood U_2 of x in A_2 such that $V_2|\overline{U}_2$ is trivial. Fix a bundle isomorphism

$$\alpha_2 : V_2|\overline{U}_2 \longrightarrow \Theta^n(\overline{U}_2).$$

Restricting to Z gives us a bundle isomorphism

$$\alpha_2|(\overline{U}_2 \cap Z) : V_2|(\overline{U}_2 \cap Z) \longrightarrow \Theta^n(\overline{U}_2 \cap Z).$$

Compose with

$$\gamma|(\overline{U}_2 \cap Z) : V_1|(\overline{U}_2 \cap Z) \longrightarrow V_2|(\overline{U}_2 \cap Z)$$

to obtain a bundle isomorphism

$$\alpha_1 : V_1|(\overline{U}_2 \cap Z) \longrightarrow \Theta^n(\overline{U}_2 \cap Z).$$

Use Proposition 1.5.19 to extend α_1 to a bundle homomorphism α_1' from V_1 to $\Theta^n(A_1)$. Next, define

$$U_1 = \{x \in A_1 : (\alpha_1')_x \text{ is an isomorphism}\}.$$

Then U_1 contains $\overline{U}_2 \cap Z$, and Proposition 1.5.20 implies that U_1 is open in A_1. By construction $\alpha_1'|(U_1 \cap U_2) = (\alpha_2\gamma)|(U_1 \cap U_2)$, so α_1' and α_2 piece together to define a bundle isomorphism

$$\alpha : (V_1 \cup_\gamma V_2)|(U_1 \cup U_2) \longrightarrow \Theta^n(U_1 \cup U_2).$$

The set $U_1 \cup U_2$ is not necessarily open in X. To correct this problem, first note that $A_2 \backslash U_2$ is closed in A_2 and A_2 is closed in X. Thus $A_2 \backslash U_2$ is closed in X. Because X is normal, there exists an open neighborhood $\widehat{U}_2$ of x in X that does not intersect $A_2 \backslash U_2$. Similarly, there exists an open neighborhood $\widehat{U}_1$ of x in X that does not intersect $A_1 \backslash U_1$. Therefore $\widehat{U}_1 \cap \widehat{U}_2$ is an open subset of X that contains x and is contained in $U_1 \cup U_2$, and thus $V_1 \cup_\gamma V_2$ is trivial over $\widehat{U}_1 \cap \widehat{U}_2$. □

1.6 Abelian monoids and the Grothendieck completion

Let X be a compact Hausdorff space. As we shall see in the next section, the sets $\text{Vect}(X)$ and $\text{Idem}(C(X))$ are each equipped with a natural notion of addition. However, $\text{Vect}(X)$ and $\text{Idem}(C(X))$ are not groups, because there are no inverses. In this section we consider a construction due to Grothendieck that remedies this problem at the potential cost of a loss of some information.

Definition 1.6.1 *An abelian monoid is a set $\mathcal{A}$ and an operation $+$ satisfying the following properties:*

(i) *The operation $+$ is associative and commutative.*
(ii) *There exists an element 0 in A with the property that $0 + a = a$ for all a in $\mathcal{A}$.*

Example 1.6.2 *Every abelian group is an abelian monoid.*

Example 1.6.3 *Let $\mathbb{Z}^+ = \{0, 1, 2, \dots\}$. Then $\mathbb{Z}^+$ is an abelian monoid under ordinary addition.*

Example 1.6.4 *Take $\mathbb{Z}^+$ and add an element ∞; denote the resulting set by $\mathbb{Z}_\infty^+$. Extend the operation of addition from $\mathbb{Z}^+$ to $\mathbb{Z}_\infty^+$ by decreeing that $k + \infty = \infty$ for every k in $\mathbb{Z}_\infty^+$. Then $\mathbb{Z}_\infty^+$ is an abelian monoid.*

Proposition 1.6.5 *Let $\mathcal{A}$ be an abelian monoid and let $\mathcal{A}^2$ be the set of ordered pairs of elements of $\mathcal{A}$. Define a relation $\sim_g$ on $\mathcal{A}^2$ by the rule $(a_1, a_2) \sim_g (b_1, b_2)$ if $a_1 + b_2 + c = a_2 + b_1 + c$ for some element c in $\mathcal{A}$. Then $\sim_g$ is an equivalence relation on $\mathcal{A}^2$, and the set $\mathcal{G}(\mathcal{A})$ of equivalence classes of $\sim_g$ is an abelian group under the operation $[(a_1, a_2)] + [(b_1, b_2)] = [(a_1 + b_1, a_2 + b_2)]$.*

Proof The reader can verify easily that $\sim_g$ is an equivalence relation on $\mathcal{A}^2$ and that addition is a well defined commutative and associative operation on $\mathcal{G}(\mathcal{A})$. The equivalence class $[(0,0)]$ serves as the identity in $\mathcal{G}(\mathcal{A})$, and the inverse of an element $[(a,b)]$ is $[(b,a)]$. Thus $\mathcal{G}(\mathcal{A})$ is an abelian group. $\qquad \square$

Definition 1.6.6 *For every abelian monoid $\mathcal{A}$, the group $\mathcal{G}(\mathcal{A})$ described in Proposition 1.6.5 is called the* Grothendieck completion *of $\mathcal{A}$.*

Instead of writing elements of $\mathcal{G}(\mathcal{A})$ in the form $[(a_1, a_2)]$, it is traditional to write them as formal differences $a_1 - a_2$. In general, it is necessary to "add on c" when defining $\mathcal{G}(\mathcal{A})$. In other words, $a_1 - a_2 = b_1 - b_2$ in $\mathcal{G}(\mathcal{A})$ does not necessarily imply that $a_1 + b_2 = a_2 + b_1$ in $\mathcal{A}$ (Exercise 1.10).

Theorem 1.6.7 *Let $\mathcal{G}(\mathcal{A})$ be the Grothendieck completion of an abelian monoid $\mathcal{A}$ and let $j : \mathcal{A} \longrightarrow \mathcal{G}(\mathcal{A})$ denote the monoid homomorphism $j(a) = a - 0$. Then for any abelian group H and monoid homomorphism $\psi : \mathcal{A} \longrightarrow H$, there is a unique group homomorphism $\widetilde{\psi} : \mathcal{G}(\mathcal{A}) \longrightarrow H$ such that $\widetilde{\psi} j = \psi$. Moreover, $\mathcal{G}(\mathcal{A})$ is unique in the following sense: suppose that there exists an abelian group $\widehat{\mathcal{G}}(\mathcal{A})$ and a monoid homomorphism $i : \mathcal{A} \longrightarrow \widehat{\mathcal{G}}(\mathcal{A})$ with the property that the image of i is a set of generators for $\widehat{\mathcal{G}}(\mathcal{A})$. Suppose further that for any abelian group H and monoid homomorphism $\psi : \mathcal{A} \longrightarrow H$, there is a unique group homomorphism $\widehat{\psi} : \widehat{\mathcal{G}}(\mathcal{A}) \longrightarrow H$ such that $\widehat{\psi} i = \psi$. Then $\mathcal{G}(\mathcal{A})$ and $\widehat{\mathcal{G}}(\mathcal{A})$ are isomorphic.*

Proof For $a_1 - a_2$ in $\mathcal{G}(\mathcal{A})$, set $\widetilde{\psi}(a_1 - a_2) = \psi(a_1) - \psi(a_2)$. We must first check that $\widetilde{\psi}$ is well defined. Suppose $a_1 - a_2 = b_1 - b_2$ in $\mathcal{G}(\mathcal{A})$. Then $a_1 + b_2 + c = a_2 + b_1 + c$ for some c in $\mathcal{A}$. Therefore $\psi(a_1) + \psi(b_2) + \psi(c) = \psi(a_2) + \psi(b_1) + \psi(c)$, and so $\widetilde{\psi}(a_1 - a_2) = \psi(a_1) - \psi(a_2) = \psi(b_1) - \psi(b_2) = \widetilde{\psi}(b_1 - b_2)$. By construction, the map $\widetilde{\psi}$ is a group homomorphism and is unique because it is determined by ψ. Finally, suppose there exists a group $\widehat{\mathcal{G}}(\mathcal{A})$ and a monoid homomorphism $i : \mathcal{A} \longrightarrow \widehat{\mathcal{G}}(\mathcal{A})$ with the properties listed in the statement of the theorem. The map i is a monoid homomorphism, so there exists a group homomorphism $\widehat{i} : \mathcal{G}(\mathcal{A}) \longrightarrow \widehat{\mathcal{G}}(\mathcal{A})$ such that $\widehat{i}j = i$. Next, because $j : \mathcal{A} \longrightarrow \mathcal{G}(\mathcal{A})$ is a monoid homomorphism, there exists a group homomorphism $\widehat{j} : \widehat{\mathcal{G}}(\mathcal{A}) \longrightarrow \mathcal{G}(\mathcal{A})$ such that $\widehat{j}i = j$. Consider the composition $\widehat{i}\widehat{j} : \widehat{\mathcal{G}}(\mathcal{A}) \longrightarrow \widehat{\mathcal{G}}(\mathcal{A})$. The image of i generates $\widehat{\mathcal{G}}(\mathcal{A})$, and therefore every element of $\widehat{\mathcal{G}}(\mathcal{A})$ can be written in the form $i(a_1) - i(a_2)$ for some a_1, a_2 in $\mathcal{A}$. Then

$$(\widehat{i}\,\widehat{j})(i(a_1) - i(a_2)) = \widehat{i}(j(a_1) - j(a_2)) = i(a_1) - i(a_2),$$

and therefore $\widehat{i}\,\widehat{j}$ is the identity map. A similar argument shows that $\widehat{j}\,\widehat{i}$ is the identity map on $\mathcal{G}(\mathcal{A})$. $\square$

Corollary 1.6.8 *Let $\mathcal{A}_1$ and $\mathcal{A}_2$ be abelian monoids and suppose that $\psi : \mathcal{A}_1 \longrightarrow \mathcal{A}_2$ is a monoid homomorphism. Let j_1 and j_2 denote the monoid homomorphisms described in the statement of Theorem 1.6.7. Then there is a unique group homomorphism $\psi' : \mathcal{G}(\mathcal{A}_1) \longrightarrow \mathcal{G}(\mathcal{A}_2)$ such that $\psi' j_1 = j_2 \psi$.*

Proof The composition $j_2 \psi : \mathcal{A}_1 \longrightarrow \mathcal{G}(\mathcal{A}_2)$ is a monoid homomorphism, and the existence and uniqueness of ψ' follow from Theorem 1.6.7. $\square$

For the remainder of the book we will omit the prime and use the same letter to denote both the monoid homomorphism from $\mathcal{A}_1$ and $\mathcal{A}_2$ and the group homomorphism from $\mathcal{G}(\mathcal{A}_1)$ and $\mathcal{G}(\mathcal{A}_2)$.

Example 1.6.9 *If $\mathcal{A}$ is an abelian monoid that is already an abelian group, then $\mathcal{G}(\mathcal{A})$ is (isomorphic to) $\mathcal{A}$.*

Example 1.6.10 *The Grothendieck completion of $\mathbb{Z}^+$ is the group of integers $\mathbb{Z}$ under addition.*

Example 1.6.11 *The Grothendieck completion of $\mathbb{Z}_\infty^+$ is the trivial group. To see this, note that for any two elements j and k in $\mathbb{Z}_\infty^+$, $j + 0 + \infty = k + 0 + \infty$, and thus $j - k = 0 - 0 = 0$.*

1.7 Vect(X) vs. Idem(C(X))

For any topological space X, the set $\mathrm{Vect}(X)$ forms an abelian monoid under internal Whitney sum. If X is a compact Hausdorff space, then we can also consider the set $\mathrm{Idem}(C(X))$ of similarity classes of idempotents over X. As we will see in a minute, $\mathrm{Idem}(C(X))$ also has the structure of an abelian monoid. The goal of this section is to prove that the Grothendieck completions of $\mathrm{Vect}(X)$ and $\mathrm{Idem}(C(X))$ are isomorphic. We begin with a lemma that we will use often in the rest of this chapter and the next.

Lemma 1.7.1 *Let X be compact Hausdorff, let m be a natural number, and let σ be an element of the permutation group S_m. Suppose that for all $1 \leq i \leq m$, we have an idempotent E_i in $\mathrm{M}(n_i, C(X))$ for some natural number n_i. Then*

$$\mathrm{diag}(\mathsf{E}_1, \mathsf{E}_2, \ldots, \mathsf{E}_m) \sim_s \mathrm{diag}(\mathsf{E}_{\sigma(1)}, \mathsf{E}_{\sigma(2)}, \ldots, \mathsf{E}_{\sigma(m)})$$

in $\mathrm{M}(n_1 + n_2 + \cdots + n_m, \mathbb{C})$.

Proof Let U be the block matrix that for each i has the n_i-by-n_i identity matrix for its $(\sigma^{-1}(i), i)$ entry and is 0 elsewhere. Then U is invertible and implements the desired similarity. $\qquad\square$

Proposition 1.7.2 *Let X be compact Hausdorff. For any two elements of $\mathrm{Idem}(C(X))$, choose representatives E in $\mathrm{M}(m, C(X))$ and F in $\mathrm{M}(n, C(X))$ and define*

$$[\mathsf{E}] + [\mathsf{F}] = [\mathrm{diag}(\mathsf{E}, \mathsf{F})].$$

Then $\mathrm{Idem}(C(X))$ is an abelian monoid.

Proof Because

$$\mathrm{diag}(\mathsf{SES}^{-1}, \mathsf{TFT}^{-1}) = \mathrm{diag}(\mathsf{S}, \mathsf{T}) \, \mathrm{diag}(\mathsf{E}, \mathsf{F}) \, \mathrm{diag}(\mathsf{S}, \mathsf{T})^{-1}$$

for all S in $\mathrm{GL}(m, C(X))$ and T in $\mathrm{GL}(n, C(X))$, we see that addition respects similarity classes. Lemma 1.7.1 gives us

$$[\mathsf{E}] + [\mathsf{F}] = [\mathrm{diag}(\mathsf{E}, 0)] + [\mathrm{diag}(\mathsf{F}, 0)] =$$
$$[\mathrm{diag}(\mathsf{E}, 0, \mathsf{F}, 0)] = [\mathrm{diag}(\mathsf{E}, \mathsf{F}, 0, 0)] = [\mathrm{diag}(\mathsf{E}, \mathsf{F})],$$

and therefore addition on $\mathrm{Idem}(C(X))$ is well defined. The zero matrix is the additive identity in $\mathrm{Idem}(C(X))$, and associativity and commutativity of addition are consequences of Lemma 1.7.1. Thus $\mathrm{Idem}(C(X))$ is an abelian monoid. $\qquad\square$

Lemma 1.7.3 *Let X be a compact Hausdorff space, let n be a natural number, and suppose E and F are idempotents in $\mathrm{M}(n, C(X))$ such that $EF = FE = 0$. Then $E + F$ is an idempotent and $[E] + [F] = [E + F]$ in Idem($C(X)$).*

Proof Because

$$(E + F)^2 = E^2 + EF + FE + F^2 = E + F,$$

the sum $E + F$ is an idempotent. We also have

$$\mathrm{diag}(E + F, 0_n) = \begin{pmatrix} E & I_n - E \\ E - I_n & E \end{pmatrix} \mathrm{diag}(E, F) \begin{pmatrix} E & E - I_n \\ I_n - E & E \end{pmatrix}$$

$$= \begin{pmatrix} E & I_n - E \\ E - I_n & E \end{pmatrix} \mathrm{diag}(E, F) \begin{pmatrix} E & I_n - E \\ E - I_n & E \end{pmatrix}^{-1},$$

and therefore

$$[E] + [F] = [\mathrm{diag}(E, F)] = [\mathrm{diag}(E + F, 0_n)] = [E + F].$$

$\square$

Definition 1.7.4 *Let X be a compact Hausdorff space, let n be a natural number, and suppose that E is an idempotent in $\mathrm{M}(n, C(X))$. For each x in X, define $\mathrm{Ran}\, E(x)$ to be the range of the matrix $E(x)$ in $\mathbb{C}^n$ and let $\mathrm{Ran}\, E$ denote the family of vector spaces $\{\mathrm{Ran}\, E(x)\}_{x \in X}$.*

Proposition 1.7.5 *Let X be a compact Hausdorff space, let n be a natural number, and suppose that E is an idempotent in $\mathrm{M}(n, C(X))$. Then $\mathrm{Ran}\, E$ is a vector bundle over X.*

Proof Fix x_0 in X and define an open set

$$U = \{x \in X : \|E(x_0) - E(x)\|_{op} < 1\}.$$

Choose x_1 in U. Proposition 1.3.4 implies that $I_n + E(x_0) - E(x_1)$ is an element of $\mathrm{GL}(n, \mathbb{C})$. Furthermore, for every v in $\mathrm{Ran}\, E(x_1)$, we have

$$(I_n + E(x_0) - E(x_1))v = (I_n + E(x_0) - E(x_1))E(x_1)v$$
$$= E(x_1)v + E(x_0)E(x_1)v - E(x_1)v$$
$$= E(x_0)E(x_1)v.$$

Thus $I_n + E(x_0) - E(x_1)$ is an injective vector space homomorphism from $\mathrm{Ran}\, E(x_1)$ to $\mathrm{Ran}\, E(x_0)$, which in turn implies that the dimension of

$\operatorname{Ran} \mathsf{E}(x_1)$ is less than or equal to the dimension of $\operatorname{Ran} \mathsf{E}(x_0)$. Similarly, $I_n + \mathsf{E}(x_1) - \mathsf{E}(x_0)$ is a vector space homomorphism from $\operatorname{Ran} \mathsf{E}(x_0)$ to $\operatorname{Ran} \mathsf{E}(x_1)$, and the dimension of $\operatorname{Ran} \mathsf{E}(x_0)$ is less than or equal to the dimension of $\operatorname{Ran} \mathsf{E}(x_1)$. Therefore $\operatorname{Ran} \mathsf{E}(x_0)$ and $\operatorname{Ran} \mathsf{E}(x_1)$ have equal dimension, and so $I_n + \mathsf{E}(x_0) - \mathsf{E}(x_1)$ is a vector space isomorphism from $\operatorname{Ran} \mathsf{E}(x_1)$ to $\operatorname{Ran} \mathsf{E}(x_0)$. As a consequence, the map $\phi : (\operatorname{Ran} \mathsf{E})|U \longrightarrow U \times \operatorname{Ran} \mathsf{E}(x_0)$ defined by

$$\phi(x, v) = (x, (I + \mathsf{E}(x_0) - \mathsf{E}(x))v)$$

is a bundle isomorphism. Our choice of x_0 was arbitrary, so $\operatorname{Ran} \mathsf{E}$ is locally trivial and hence a vector bundle over X.   □

Proposition 1.7.6 *Suppose X is compact Hausdorff and let E and F be idempotents over X. Then $[\mathsf{E}] = [\mathsf{F}]$ in $\operatorname{Idem}(C(X))$ if and only if $[\operatorname{Ran} \mathsf{E}] = [\operatorname{Ran} \mathsf{F}]$ in $\operatorname{Vect}(X)$.*

Proof Because $[\operatorname{Ran} \mathsf{E}] = [\operatorname{Ran}(\operatorname{diag}(\mathsf{E}, 0))]$ in $\operatorname{Idem}(X)$ for any idempotent E over X, we may choose a natural number n large enough so that both E and F are in $M(n, C(X))$ and that $\mathsf{F} = \mathsf{SES}^{-1}$ for some S in $\operatorname{GL}(n, C(X))$. Every element of $\operatorname{Ran} \mathsf{E}$ can be written in the form $\mathsf{E}\vec{f}$, where $\vec{f}$ is in $(C(X))^n$; define a linear map $\gamma : \operatorname{Ran} \mathsf{E} \longrightarrow \operatorname{Ran} \mathsf{F}$ by

$$\gamma(\mathsf{E}\vec{f}) = \mathsf{SE}\vec{f} = \mathsf{FS}\vec{f}.$$

The inverse map is defined by the formula

$$\gamma^{-1}(\mathsf{F}\vec{f}) = \mathsf{S}^{-1}\mathsf{F}\vec{f} = \mathsf{ES}^{-1}\vec{f},$$

so $\operatorname{Ran} \mathsf{E}$ is isomorphic to $\operatorname{Ran} \mathsf{F}$ and thus $[\operatorname{Ran} \mathsf{E}] = [\operatorname{Ran} \mathsf{F}]$ in $\operatorname{Vect}(X)$.

Conversely, suppose that $[\operatorname{Ran} \mathsf{E}] = [\operatorname{Ran} \mathsf{F}]$. Choose an isomorphism $\delta : \operatorname{Ran} \mathsf{E} \longrightarrow \operatorname{Ran} \mathsf{F}$, and for each $\vec{f}$ in $(C(X))^n$, define $\mathsf{A}\vec{f} = \delta(\mathsf{E}\vec{f})$ and $\mathsf{B}\vec{f} = \delta^{-1}(\mathsf{F}\vec{f})$. Then

$$\mathsf{AB}\vec{f} = \mathsf{A}\big(\delta^{-1}(\mathsf{F}\vec{f})\big) = \delta\big(\mathsf{E}(\delta^{-1}(\mathsf{F}\vec{f}))\big).$$

But $\delta^{-1}(\mathsf{F}\vec{f})$ is in the range of E, so

$$\mathsf{AB}\vec{f} = \delta\big(\mathsf{E}(\delta^{-1}(\mathsf{F}\vec{f}))\big) = \delta\delta^{-1}(\mathsf{F}\vec{f}) = \mathsf{F}\vec{f}.$$

Thus $\mathsf{AB} = \mathsf{F}$. Similar computations show that $\mathsf{BA} = \mathsf{E}$, $\mathsf{EB} = \mathsf{B} = \mathsf{BF}$, and $\mathsf{FA} = \mathsf{A} = \mathsf{AE}$. Define

$$T = \begin{pmatrix} \mathsf{A} & I_n - \mathsf{F} \\ I_n - \mathsf{E} & \mathsf{B} \end{pmatrix}.$$

Then T is invertible with inverse

$$T^{-1} = \begin{pmatrix} B & I_n - E \\ I_n - F & A \end{pmatrix},$$

and

$$T \operatorname{diag}(E, 0_n) T^{-1} = \operatorname{diag}(F, 0_n).$$

Therefore $[E] = [F]$ in Idem($C(X)$). $\square$

One consequence of Proposition 1.7.6 is that $[E] \mapsto [\operatorname{Ran} E]$ is an injection from Idem($C(X)$) to Vect(X) for every compact Hausdorff space X. Our next goal is to show that this map is in fact a bijection.

Definition 1.7.7 *Let V be a vector bundle over a compact Hausdorff space X, endow the set $V \times V$ with the product topology, and equip $\bigcup_{x \in X}(V_x \times V_x)$ with the subspace topology it inherits from $V \times V$. A Hermitian metric on V is a continuous function*

$$\langle \cdot, \cdot \rangle : \bigcup_{x \in X} V_x \times V_x \longrightarrow \mathbb{C}$$

with the property that for each x in X, the restriction of $\langle \cdot, \cdot \rangle$ to $V_x \times V_x$ is a complex inner product. For each v in V, we define $\|v\|_{in} = \sqrt{\langle v, v \rangle}$.

For each natural number n and compact Hausdorff space X, we can define a Hermitian metric on the trivial bundle $\Theta^n(X)$ by taking the standard inner product on each fiber $\Theta^n(X)_x \cong \mathbb{C}^n$; we will call this metric the *standard Hermitian metric* on $\Theta^n(X)$.

Lemma 1.7.8 *Let V and W be vector bundles over a compact Hausdorff space X and suppose that $\alpha : V \longrightarrow W$ is a bundle homomorphism with the property that α_x is a vector space isomorphism for every x in X. Then α is a bundle isomorphism.*

Proof We may and do assume that X is connected. The map α is a bijection, so α^{-1} exists; we must show that α^{-1} is continuous. Choose a finite open cover $\{U_1, U_2, \ldots, U_m\}$ of X with the property that $V|_{\overline{U}_k}$ and $W|_{\overline{U}_k}$ are both trivial for each $1 \le k \le m$. For each k, take $\alpha_k : V|_{\overline{U}_k} \longrightarrow W|_{\overline{U}_k}$ to be the restriction of α. Then we need only show that each α_k^{-1} is continuous.

Let n denote the rank of V and W and identify $V|_{\overline{U}_k}$ and $W|_{\overline{U}_k}$ with $\Theta^n(\overline{U}_k)$. We consider α_k to be an element of Hom($\Theta^n(\overline{U}_k), \Theta^n(\overline{U}_k)$) and apply Lemma 1.5.17 to obtain a continuous function $f_k : \overline{U}_k \longrightarrow \mathrm{M}(n, \mathbb{C})$

such that $\alpha_k(x, \vec{z}) = f_k(x)\vec{z}$ for all $(x, \vec{z})$ in $\Theta^n(\overline{U}_k)$. Because $\alpha_k(x)$ is invertible for every x in U_k, the range of f_k is contained in $\mathrm{GL}(n, \mathbb{C})$.

Next, by Lemma 1.5.17 there exists an element F_k in $\mathrm{M}(n, C(\overline{U}_k))$ such that $F_k(x) = f_k(x)$ for every x in $\overline{U}_k$. In fact, F_k is an element of $\mathrm{GL}(n, C(\overline{U}_k))$ because f_k is an element of $\mathrm{GL}(n, \mathbb{C})$. We know from Proposition 1.3.8 that $\mathrm{GL}(n, C(\overline{U}_k))$ is a topological group, whence F_k^{-1} is in $\mathrm{GL}(n, C(\overline{U}_k))$.

Applying Lemma 1.5.17 to F_k^{-1}, we see that $f_k^{-1} : \overline{U}_k \longrightarrow \mathrm{GL}(n, \mathbb{C})$ is continuous, and finally Lemma 1.5.17 yields that α_k^{-1} is continuous. $\qquad\square$

Proposition 1.7.9 *Let V be a vector bundle over a compact Hausdorff space X. Then V is isomorphic to a subbundle of $\Theta^N(X)$ for some natural number N.*

Proof Let $X_1, X_2, \ldots, X_m$ be the connected components of X, and suppose for $1 \leq k \leq m$, the bundle $V|X_k$ is isomorphic to a subbundle of a trivial bundle of rank N_k. If we take $N = N_1 + N_2 + \cdots + N_m$, then V is isomorphic to a subbundle of $\Theta^N(X)$. Thus we may assume that X is connected.

Choose a finite open cover $\mathcal{U} = \{U_1, U_2, \ldots, U_l\}$ of X with the property that $V|U_k$ is trivial for $1 \leq k \leq l$. Let n be the rank of V and for each k, let ϕ_k be a bundle isomorphism from $V|U_k$ to $\Theta^n(U_k)$. Next, define $q_k : \Theta^n(U_k) \longrightarrow \mathbb{C}^n$ as $q_k(x, \vec{z}) = \vec{z}$ for every x in X and $\vec{z}$ in $\mathbb{C}^n$. Choose a partition of unity $\{p_1, p_2, \ldots, p_l\}$ subordinate to $\mathcal{U}$, set $N = nl$, and define $\Phi : V \longrightarrow \mathbb{C}^n \oplus \cdots \oplus \mathbb{C}^n = \mathbb{C}^{ln}$ by the formula

$$\Phi(v) = (p_1(\pi(v))q_1(\phi_1(v)), p_2(\pi(v))q_2(\phi_2(v)), \ldots, p_l(\pi(v))q_l(\phi_l(v))).$$

Then $\phi(v) = (\pi(v), \Phi(v))$ defines a bundle homomorphism from V into $\Theta^N(X)$. Moreover, the map ϕ is injective, so the desired result follows from Lemma 1.7.8. $\qquad\square$

Corollary 1.7.10 *Every vector bundle over a compact Hausdorff space admits a Hermitian metric.*

Proof If V is a vector bundle over a compact Hausdorff space X, we can by Proposition 1.7.9 imbed V as a subbundle of $\Theta^N(X)$ for some sufficiently large natural number N. Then the standard Hermitian metric on $\Theta^N(X)$ restricts to give a metric on V. $\qquad\square$

Lemma 1.7.11 *Let X be a compact Hausdorff space, let N be a natural number, and suppose that V is a subbundle of $\Theta^N(X)$. Equip $\Theta^N(X)$ with its standard Hermitian metric and for each x in X, let $\mathsf{E}(x)$ be the orthogonal projection of $\Theta^N(X)_x$ onto V_x. Then the collection $\{\mathsf{E}(x)\}$ defines an idempotent E in $\mathsf{M}(N, C(X))$.*

Proof Thanks to Lemma 1.5.17, we need only show that each x_0 in X has an open neighborhood U with the property that the map $x \mapsto \mathsf{E}(x)$ is continuous when restricted to U. Fix x_0 and choose U to be a connected open neighborhood of x_0 over which V is trivial. Let n be the rank of V over U and fix a bundle isomorphism ϕ from $\Theta^n(U)$ to $V|U$. Next, for $1 \leq k \leq n$ define $s_k : U \longrightarrow \Theta^n(U)$ by the formula $s_k(x) = (x, e_k)$. Then for each x in U the set $\{\phi(s_1(x)), \phi(s_2(x)), \ldots, \phi(s_n(x))\}$ is a vector space basis for V_x. Restrict the standard Hermitian metric on $\Theta^N(X)$ to define a Hermitian metric on V. From Proposition 1.1.2, we obtain an orthogonal basis of V_x via the formula

$$s_k'(x) = \phi(s_k(x)) - \sum_{i=1}^{k-1} \frac{\langle \phi(s_k(x)), s_i'(x) \rangle}{\langle s_i'(x), s_i'(x) \rangle} s_i'(x)$$

for each $1 \leq k \leq n$ and each x in U. Then

$$\left\{ \frac{s_1'(x)}{\|s_1'(x)\|_{in}}, \frac{s_2'(x)}{\|s_2'(x)\|_{in}}, \ldots, \frac{s_n'(x)}{\|s_n'(x)\|_{in}} \right\}$$

is an orthonormal basis of V_x. These basis elements vary continuously in x. Moreover, we can combine Proposition 1.1.6 and Definition 1.1.7 to write down the formula

$$\mathsf{E}(x)\vec{z} = \sum_{k=1}^{n} \left\langle \phi(x, \vec{z}), \frac{s_k'(x)}{\|s_k'(x)\|_{in}} \right\rangle \frac{s_1'(x)}{\|s_1'(x)\|_{in}},$$

which is a continuous function of x. □

Proposition 1.7.12 *Let V be a vector bundle over a compact Hausdorff space X. Then there exists a vector bundle $V^\perp$ over X such that $V \oplus V^\perp$ is isomorphic to $\Theta^N(X)$ for some natural number N.*

Proof Choose N large enough so that V is (isomorphic to) a subbundle of $\Theta^N(X)$. For each x in X, let $\mathsf{E}(x)$ be the orthogonal projection of $\Theta^N(X)_x$ onto V_x. By Lemma 1.7.11 this family of orthogonal projections defines an idempotent E in $\mathsf{M}(N, C(X))$. Define $V^\perp = \mathrm{Ran}(I_N - \mathsf{E})$.

Then

$$V \oplus V^{\perp} \cong \operatorname{Ran} \mathsf{E} \oplus \operatorname{Ran}(I_N - \mathsf{E}) = \operatorname{Ran} I_N = \Theta^N(X).$$

$\square$

Corollary 1.7.13 *Let V be a vector bundle over a compact Hausdorff space X. Then $V = \operatorname{Ran} \mathsf{E}$ for some idempotent E over X.*

Proof Imbed V as a subbundle of $\Theta^N(X)$ for some sufficiently large natural number N and for each x in X, let $\mathsf{E}(x)$ be the orthogonal projection of $\Theta^N(X)_x$ onto V_x. Then apply Lemma 1.7.11. $\square$

Theorem 1.7.14 *Let X be a compact Hausdorff space. Then the abelian monoids $\operatorname{Idem}(C(X))$ and $\operatorname{Vect}(X)$ are isomorphic.*

Proof Define $\psi : \operatorname{Idem}(C(X)) \longrightarrow \operatorname{Vect}(X)$ as $\psi[\mathsf{E}] = [\operatorname{Ran} \mathsf{E}]$. We know that ψ is well defined and injective from Proposition 1.7.6. Moreover, ψ is a monoid homomorphism, and Corollary 1.7.13 yields that ψ is onto. $\square$

We end this section by proving a result we will need in Chapter 3.

Proposition 1.7.15 *Let X be a compact Hausdorff space, let V be a vector bundle over X, and suppose that W is a subbundle of V. Then there exists a vector subbundle W' of V with the property that $W \oplus W'$ is isomorphic to V.*

Proof Proposition 1.7.12 implies that we can imbed V into a trivial bundle $\Theta^N(X)$ for some natural number N. Use Corollary 1.7.13 to produce idempotents E and F in $\mathrm{M}(N, C(X))$ such that $V = \operatorname{Ran} \mathsf{E}$ and $W = \operatorname{Ran} \mathsf{F}$. Because W is a subbundle of V, we have the equalities

$$\mathsf{EF} = \mathsf{FE} = \mathsf{F}.$$

Therefore

$$(\mathsf{E} - \mathsf{F})^2 = \mathsf{E}^2 - \mathsf{EF} - \mathsf{FE} + \mathsf{F}^2 = \mathsf{E} - \mathsf{F} - \mathsf{F} + \mathsf{F} = \mathsf{E} - \mathsf{F},$$

and thus $\mathsf{E} - \mathsf{F}$ is an idempotent. Set $W' = \operatorname{Ran}(\mathsf{E} - \mathsf{F})$ to obtain the desired vector bundle. $\square$

1.8 Some homological algebra

Definition 1.8.1 *Let $\mathcal{G}_1$, $\mathcal{G}_2$, and $\mathcal{G}_3$ be abelian groups. A pair of group homomorphisms*

$$\mathcal{G}_1 \xrightarrow{\phi_1} \mathcal{G}_2 \xrightarrow{\phi_2} \mathcal{G}_3$$

is exact *at $\mathcal{G}_2$ if the image of ϕ_1 equals the kernel of ϕ_2. More generally, a finite sequence*

$$\mathcal{G}_1 \xrightarrow{\phi_1} \mathcal{G}_2 \xrightarrow{\phi_2} \cdots \xrightarrow{\phi_{n-2}} \mathcal{G}_{n-1} \xrightarrow{\phi_{n-1}} \mathcal{G}_n$$

is an exact sequence *if it is exact at $\mathcal{G}_2$, $\mathcal{G}_3$, ..., $\mathcal{G}_{n-1}$; the notion of exactness is undefined at $\mathcal{G}_1$ and $\mathcal{G}_n$.*

Let 0 *denote the trivial group. An exact sequence of the form*

$$0 \longrightarrow \mathcal{G}_1 \xrightarrow{\phi_1} \mathcal{G}_2 \xrightarrow{\phi_2} \mathcal{G}_3 \longrightarrow 0$$

is called a short exact sequence.

For the short exact sequence in Definition 1.8.1, we note that the map ϕ_1 is injective and the map ϕ_2 is surjective. Furthermore, $\phi_1(\mathcal{G}_1)$ is a normal subgroup of $\mathcal{G}_2$, and the quotient group is isomorphic to $\mathcal{G}_3$.

Example 1.8.2 *Let $\mathcal{G}$ and $\mathcal{H}$ be any two abelian groups. Define $i : \mathcal{G} \longrightarrow \mathcal{G} \oplus \mathcal{H}$ by $i(g) = (g, 0)$ and define $p : \mathcal{G} \oplus \mathcal{H} \longrightarrow \mathcal{H}$ as $p(g, h) = h$. Then*

$$0 \longrightarrow \mathcal{G} \xrightarrow{i} \mathcal{G} \oplus \mathcal{H} \xrightarrow{p} \mathcal{H} \longrightarrow 0$$

is a short exact sequence.

Example 1.8.3 *Define $\phi_1 : \mathbb{Z}_2 \longrightarrow \mathbb{Z}_4$ as $\phi_1(0) = 0$ and $\phi_1(1) = 2$, and define ϕ_2 to be the homomorphism that maps 0 and 2 to 0, and maps 1 and 3 to 1. Then the sequence*

$$0 \longrightarrow \mathbb{Z}_2 \xrightarrow{\phi_1} \mathbb{Z}_4 \xrightarrow{\phi_2} \mathbb{Z}_2 \longrightarrow 0$$

is exact.

Definition 1.8.4 *A short exact sequence*

$$0 \longrightarrow \mathcal{G}_1 \xrightarrow{\phi_1} \mathcal{G}_2 \xrightarrow{\phi_2} \mathcal{G}_3 \longrightarrow 0$$

is split exact *if there exists a group homomorphism $\psi : \mathcal{G}_3 \longrightarrow \mathcal{G}_2$ such that $\phi_2\psi$ is the identity map on $\mathcal{G}_3$; we call such a homomorphism ψ a* splitting map. *We write split exact sequences using the notation*

$$0 \longrightarrow \mathcal{G}_1 \xrightarrow{\phi_1} \mathcal{G}_2 \underset{\psi}{\xrightarrow{\phi_2}} \mathcal{G}_3 \longrightarrow 0.$$

Example 1.8.5 *The exact sequence*

$$0 \longrightarrow \mathcal{G} \xrightarrow{i} \mathcal{G} \oplus \mathcal{H} \xrightarrow{p} \mathcal{H} \longrightarrow 0$$

from Example 1.8.2 is split exact because the homomorphism $\psi : \mathcal{H} \longrightarrow \mathcal{G} \oplus \mathcal{H}$ given by the formula $\psi(h) = (0, h)$ is a splitting map.

Example 1.8.6 *Suppose that*

$$0 \longrightarrow \mathcal{G}_1 \xrightarrow{\phi_1} \mathcal{G}_2 \xrightarrow{\phi_2} \mathbb{Z}^n \longrightarrow 0$$

is a short exact sequence of abelian groups. For each $1 \leq k \leq n$, let e_k denote the element of $\mathbb{Z}^k$ that equals 1 in the kth entry and is 0 in the other entries. We can then uniquely write each element of $\mathbb{Z}^k$ in the form $m_1e_1 + m_2e_2 + \cdots + m_ne_n$ for some integers $m_1, m_2, \ldots, m_n$.

For each e_k, choose an element a_k in $\mathcal{G}_2$ such that $\phi_2(a_k) = e_k$. Then the homomorphism

$$\psi(m_1e_1 + m_2e_2 + \cdots + m_ne_n) = m_1a_1 + m_2a_2 + \cdots + m_na_n$$

is a splitting map, and thus the short exact sequence is split exact.

Proposition 1.8.7 *Let*

$$0 \longrightarrow \mathcal{G}_1 \xrightarrow{\phi_1} \mathcal{G}_2 \underset{\psi_3}{\xrightarrow{\phi_2}} \mathcal{G}_3 \longrightarrow 0$$

be a split exact sequence of abelian groups. Then $\phi_1 + \psi_3 : \mathcal{G}_1 \oplus \mathcal{G}_3 \longrightarrow \mathcal{G}_2$ is an isomorphism, and there exists a homomorphism $\psi_2 : \mathcal{G}_2 \longrightarrow \mathcal{G}_1$ with the property that $(\phi_1 + \psi_3)^{-1} = \psi_2 \oplus \phi_2$.

Proof Take g_2 in $\mathcal{G}_2$. Then

$$\phi_2(g_2 - \psi_3\phi_2(g_2)) = \phi_2(g_2) - \phi_2\psi_3\phi_2(g_2) = \phi_2(g_2) - \phi_2(g_2) = 0.$$

Thus $g_2 - \psi_3\phi_2(g_2)$ is in the image of ϕ_1; define $\psi_2(g_2)$ to be the unique element of $\mathcal{G}_1$ with the property that $\phi_1\psi_2(g_2) = g_2 - \psi_3\phi_2(g_2)$. The reader can readily check that ψ_2 is a homomorphism. The definition of ψ_2 and exactness at $\mathcal{G}_2$ together imply

$$\phi_1\phi_2\phi_1(g_1) = \phi_1(g_1) - \psi_3\phi_2\phi_1(g_1) = \phi_1(g_1)$$

for all g_1 in $\mathcal{G}_1$. The injectivity of ϕ_1 then yields that $\psi_2\phi_1$ is the identity on $\mathcal{G}_1$. Next, for each g_3 in $\mathcal{G}_3$, the definition of ψ_2 gives us

$$\phi_1\psi_2\psi_3(g_3) = \psi_3(g_3) - \psi_3\phi_2\psi_3(g_3) = \psi_3(g_3) - \psi_3(g_3) = 0,$$

and the injectivity of ϕ_1 then implies that $\psi_2\psi_3 = 0$. Therefore

$$(\psi_2 \oplus \phi_2)(\phi_1 + \psi_3) = \psi_2\phi_1 \oplus \phi_2\psi_3$$

is the identity map on $\mathcal{G}_1 \oplus \mathcal{G}_2$, and $\psi_2 \oplus \phi_2$ is a surjection.

Finally, we show that $\psi_2 \oplus \phi_2$ is an injection, which will imply that $\phi_1 + \psi_3$ is its inverse. Suppose that $(\psi_2 \oplus \phi_2)(g_2) = 0$. Then $\phi_2(g_2) = 0$, whence $g_2 = \phi_1(g_1)$ for some g_1 in $\mathcal{G}_1$. Moreover, we have $0 = \psi_2(g_2) = \psi_2\phi_1(g_1) = g_1$, whence g_2 is zero. $\qquad\square$

Definition 1.8.8 *A square diagram of sets and functions*

commutes *if* $\gamma\alpha = \delta\beta$. *More generally, we say a diagram is commutative if each square in the diagram commutes.*

We state and prove two homological algebra results that we will find useful throughout the book; these propositions are proved by what is often called a "diagram chase."

Proposition 1.8.9 *Suppose that the diagram*

$$0 \longrightarrow \mathcal{G}_1 \xrightarrow{\ \phi_1\ } \mathcal{G}_2 \xrightarrow{\ \phi_2\ } \mathcal{G}_3$$
$$\downarrow{\alpha_2} \qquad \downarrow{\alpha_3}$$
$$0 \longrightarrow \mathcal{H}_1 \xrightarrow{\ \psi_1\ } \mathcal{H}_2 \xrightarrow{\ \psi_2\ } \mathcal{H}_3.$$

of abelian groups is commutative and has exact rows. Then there exists a unique group homomorphism $\alpha_1 : \mathcal{G}_1 \longrightarrow \mathcal{H}_1$ *that makes the diagram*

$$
\begin{array}{ccccccc}
0 & \longrightarrow & \mathcal{G}_1 & \xrightarrow{\phi_1} & \mathcal{G}_2 & \xrightarrow{\phi_2} & \mathcal{G}_3 \\
 & & \downarrow{\alpha_1} & & \downarrow{\alpha_2} & & \downarrow{\alpha_3} \\
0 & \longrightarrow & \mathcal{H}_1 & \xrightarrow{\psi_1} & \mathcal{H}_2 & \xrightarrow{\psi_2} & \mathcal{H}_3
\end{array}
$$

commute. Furthermore, if α_2 *and* α_3 *are isomorphisms, so is* α_1.

Proof For each g_1 in $\mathcal{G}_1$, the element $\alpha_2 \phi_1(g_1)$ is in $\mathcal{H}_2$, and

$$\psi_2 \alpha_2 \phi_1(g_1) = \alpha_3 \phi_2 \phi_1(g_1) = 0.$$

Thus $\alpha_2 \phi_1(g_1)$ is in the image of ψ_1, and therefore there exists an element h_1 in $\mathcal{H}_1$ with the property that $\psi_1(h_1) = \alpha_2 \phi_1(g_1)$. Moreover, exactness tells us that ψ_1 is an injection, whence h_1 is unique. Set $\alpha_1(g_1) = h_1$; an easy computation shows that α_1 is a group homomorphism.

Now suppose α_2 and α_3 are isomorphisms. If $\alpha_1(g_1) = 0$ for some g_1 in $\mathcal{G}_1$, the commutativity of the diagram gives us

$$0 = \psi_1 \alpha_1(g_1) = \alpha_2 \phi_1(g_1).$$

We know that $\phi_1(g_1) = 0$ because α_2 is an isomorphism, and $g_1 = 0$ because ϕ_1 is injective. Therefore α_1 is injective.

Next, take $\tilde{h}_1$ in $\mathcal{H}_1$ Then

$$\alpha_3 \phi_2 \alpha_2^{-1} \psi_1(\tilde{h}_1) = \psi_2 \alpha_2 \alpha_2^{-1} \psi_1(\tilde{h}_1) = \psi_2 \psi_1(\tilde{h}_1) = 0,$$

and because α_3 is an isomorphism, we have $\phi_2 \alpha_2^{-1} \psi_1(\tilde{h}_1) = 0$. The top sequence is exact at $\mathcal{G}_2$, whence $\alpha_2^{-1} \psi_1(\tilde{h}_1) = \phi_1(\tilde{g}_1)$ for some $\tilde{g}_1$ in $\mathcal{G}_1$. Moreover, $\psi_1 \alpha_1(\tilde{g}_1) = \alpha_2 \phi_1(\tilde{g}_1) = \psi_1(\tilde{h}_1)$, and thus $\alpha(\tilde{g}_1) - \tilde{h}_1$ is in the kernel of ψ_1. But ψ_1 is injective, so $\alpha(\tilde{g}_1) = \tilde{h}_1$, and therefore α_1 is surjective. $\qquad\qquad\square$

Proposition 1.8.10 (Five lemma) *Suppose that*

$$
\begin{array}{ccccccccc}
\mathcal{G}_1 & \xrightarrow{\phi_1} & \mathcal{G}_2 & \xrightarrow{\phi_2} & \mathcal{G}_3 & \xrightarrow{\phi_3} & \mathcal{G}_4 & \xrightarrow{\phi_4} & \mathcal{G}_5 \\
\downarrow{\alpha_1} & & \downarrow{\alpha_2} & & \downarrow{\alpha_3} & & \downarrow{\alpha_4} & & \downarrow{\alpha_5} \\
\mathcal{H}_1 & \xrightarrow{\psi_1} & \mathcal{H}_2 & \xrightarrow{\psi_2} & \mathcal{H}_3 & \xrightarrow{\psi_3} & \mathcal{H}_4 & \xrightarrow{\psi_4} & \mathcal{H}_5
\end{array}
$$

is a diagram of abelian groups that is commutative and has exact rows. Further suppose that α_1, α_2, α_4, and α_5 are isomorphisms. Then α_3 is an isomorphism.

Proof Suppose that $\alpha_3(g_3) = 0$. Then $0 = \psi_3\alpha_3(g_3) = \alpha_4\phi_3(g_3)$, and because α_4 is an isomorphism, the element g_3 is in the kernel of ϕ_3, which by exactness means that g_3 is in the image of ϕ_2. Choose g_2 in $\mathcal{G}_2$ so that $\phi_2(g_2) = g_3$. Then $0 = \alpha_3\phi_2(g_2) = \psi_2\alpha_2(g_2)$. Therefore there exists h_1 in $\mathcal{H}_1$ such that $\psi_1(h_1) = \alpha_2(g_2)$. Let $g_1 = \alpha_1^{-1}(h_1)$. Then

$$\phi_1(g_1) = \alpha_2^{-1}\psi_1\alpha_1(g_1) = \alpha_2^{-1}\psi_1(h_1) = \alpha_2^{-1}\alpha_2(g_2) = g_2,$$

and thus $\phi_2\phi_1(g_1) = g_3$. But exactness implies that $\phi_2\phi_1$ is the zero homomorphism, so $g_3 = 0$, and therefore α is an injection.

To show that α_3 is surjective, choose $\tilde{h}_3$ in $\mathcal{H}_3$ and let $\tilde{g}_4 = \alpha_4^{-1}\psi_3(\tilde{h}_3)$ in $\mathcal{G}_4$. Then

$$0 = \psi_4\psi_3(\tilde{h}_3) = \psi_4\alpha_4(\tilde{g}_4) = \alpha_5\phi_4(\tilde{g}_4),$$

and therefore $\phi_4(\tilde{g}_4) = 0$. By exactness, there exists an element $\tilde{g}_3$ in $\mathcal{G}_3$ such that $\phi_3(\tilde{g}_3) = \tilde{g}_4$. The string of equalities

$$\psi_3\alpha_3(\tilde{g}_3) = \alpha_4\phi_3(\tilde{g}_3) = \alpha_4(\tilde{g}_4) = \psi_3(\tilde{h}_3)$$

shows that $\alpha_3(\tilde{g}_3) - \tilde{h}_3$ is in the kernel of ψ_3, and hence also in the image of ψ_2. Choose $\tilde{h}_2$ in $\mathcal{H}_2$ with the property that $\psi_2(\tilde{h}_2) = \alpha_3(\tilde{g}_3) - \tilde{h}_3$. Then

$$\alpha_3\left(\tilde{g}_3 - \phi_2\alpha_2^{-1}(\tilde{h}_2)\right) = \alpha_3(\tilde{g}_3) - \alpha_3\phi_2\alpha_2^{-1}(\tilde{h}_2)$$
$$= \alpha_3(\tilde{g}_3) - \psi_2\alpha_2\alpha_2^{-1}(\tilde{h}_2) = \alpha_3(\tilde{g}_3) - \psi_2(\tilde{h}_2) = \tilde{h}_3.$$

$\square$

1.9 A very brief introduction to category theory

Many results in algebraic topology, and K-theory in particular, are most cleanly expressed in the language of *category theory*.

Definition 1.9.1 *A* category *consists of:*

  (i) *a class of* objects;

  (ii) *for each ordered pair of objects (A, B), a set* Hom(A, B) *whose elements are called* morphisms;

(iii) *for each ordered triple (A, B, C) of objects, a composition law*

$$\circ : \mathrm{Hom}(A, B) \times \mathrm{Hom}(B, C) \longrightarrow \mathrm{Hom}(A, C)$$

satisfying the following properties:

(a) *If α is in $\mathrm{Hom}(A, B)$, β is in $\mathrm{Hom}(B, C)$, and γ is in $\mathrm{Hom}(C, D)$, then $\gamma \circ (\beta \circ \alpha) = (\gamma \circ \beta) \circ \alpha$.*

(b) *For every object A, there is an element 1_A in $\mathrm{Hom}(A, A)$ with the property that for any two objects B and Z and morphisms α in $\mathrm{Hom}(A, B)$ and ζ in $\mathrm{Hom}(Z, A)$, we have $1_A \circ \alpha = \alpha$ and $\zeta \circ 1_A = \zeta$.*

The objects of a category form a *class*, not necessarily a *set*; i.e., the collection of objects in a category may be "too big" to be a set. This is in fact the situation in the most commonly occurring examples of categories. In any event, the collection $\mathrm{Hom}(A, B)$ of morphisms from an object A to an object B is always required to be a set.

Example 1.9.2 *For every pair of sets A and B, let $\mathrm{Hom}(A, B)$ be the set of functions from A to B. Then the class of sets with functions for morphisms is a category.*

Example 1.9.3 *The class of all abelian monoids with monoid homomorphisms as morphisms forms a category.*

Example 1.9.4 *The class of all abelian groups with group homomorphisms as morphisms defines a category.*

Example 1.9.5 *The class of all topological spaces with continuous functions as morphisms is a category.*

Example 1.9.6 *The class of all compact Hausdorff spaces with continuous functions as morphisms forms a category.*

If there is no possibility of confusion, we will usually only mention the objects of a category, with the morphisms being understood from context. So, for example, we will refer to the category of topological spaces, the category of abelian groups, etc.

Definition 1.9.7 *Let $\mathfrak{A}$ and $\mathfrak{B}$ be categories. A covariant functor F from $\mathfrak{A}$ to $\mathfrak{B}$ is:*

(i) *a function that assigns to each object A in $\mathfrak{A}$ an object $F(A)$ in $\mathfrak{B}$, and*

(ii) *given objects A and A' in $\mathfrak{A}$, a function that assigns to each morphism $\alpha \in \mathrm{Hom}(A, A')$ a morphism $F(\alpha) \in \mathrm{Hom}(F(A), F(A'))$.*

In addition, for all objects A, A', and A'' in $\mathfrak{A}$ and all morphisms α in $\mathrm{Hom}(A, A')$ and α' in $\mathrm{Hom}(A', A'')$, we require that

$$F(1_A) = 1_{F(A)}$$

and

$$F(\alpha' \circ \alpha) = F(\alpha') \circ F(\alpha).$$

Definition 1.9.8 *Let $\mathfrak{A}$ and $\mathfrak{B}$ be categories. A* contravariant functor *F from $\mathfrak{A}$ to $\mathfrak{B}$ is:*

(i) *a function that assigns to each object A in $\mathfrak{A}$ an object $F(A)$ in $\mathfrak{B}$, and*

(ii) *given objects A and A' in $\mathfrak{A}$, a function that assigns to each morphism $\alpha \in \mathrm{Hom}(A, A')$ a morphism $F(\alpha) \in \mathrm{Hom}(F(A'), F(A))$.*

In addition, for all objects A, A', and A'' in $\mathfrak{A}$ and all morphisms $\alpha \in \mathrm{Hom}(A, A')$ and $\alpha' \in \mathrm{Hom}(A', A'')$, we require that

$$F(1_A) = 1_{F(A)}$$

and

$$F(\alpha' \circ \alpha) = F(\alpha) \circ F(\alpha').$$

Example 1.9.9 *For each topological space, consider it as a set, and for each continuous map between topological spaces, consider it as a map between sets. This process determines a covariant functor from the category of topological spaces to the category of sets. This is an example of a* forgetful *functor.*

Example 1.9.10 *Proposition 1.6.5 and Theorem 1.6.7 together imply that the Grothendieck completion of an abelian monoid defines a convariant functor from the category of abelian monoids to the category of abelian groups.*

Example 1.9.11 *Let X and X' be topological spaces. Then $\mathrm{Vect}(X)$ and $\mathrm{Vect}(X')$ are abelian monoids. Furthermore, given a continuous function $\phi : X \longrightarrow X'$, the formula $(\mathrm{Vect}(\phi))(V) = \phi^* V$ defines a monoid homomorphism $\mathrm{Vect}(\phi) : \mathrm{Vect}(X') \longrightarrow \mathrm{Vect}(X)$. Properties of* *pullback imply that* Vect *is a contravariant functor from the category of topological spaces to the category of abelian monoids.*

Example 1.9.12 *Consider the category* $\mathfrak{C}$ *whose objects are the Banach algebras* $C(X)$, *where* X *ranges over all compact Hausdorff spaces, and whose morphisms are continuous algebra homomorphisms. To construct a contravariant functor* C *from the category of compact Hausdorff spaces to* $\mathfrak{C}$, *we assign to each compact Hausdorff space* X *the Banach algebra* $C(X)$. *Given a continuous function* $\phi : X \longrightarrow X'$, *define* $C(\phi) : C(X') \longrightarrow C(X)$ *by the formula* $(C(\phi))f = f \circ \phi$ *for all* f *in* $C(X')$. *Then* $C(\mathrm{id}_X) = \mathrm{id}_{C(X)}$ *for all compact Hausdorff spaces* X. *Moreover, suppose that* X'' *is compact Hausdorff and that* $\phi' : X' \longrightarrow X''$ *is continuous. Because*

$$(C(\phi' \circ \phi))g = g \circ \phi' \circ \phi = (C(\phi) \circ C(\phi'))g$$

for all g *in* $C(X)$, *we see that* C *is a contravariant functor.*

Example 1.9.13 *We generalize Example 1.9.12 to matrices. Let* $\mathfrak{MC}$ *be the category whose objects are* $\mathrm{M}(C(X))$, *where* X *ranges over all compact Hausdorff spaces, and whose morphisms are continuous algebra homomorphisms. Associate to each compact Hausdorff space* X *the algebra* $\mathrm{M}(C(X))$. *Next, given a compact Hausdorff space* X', *write each element* A *of* $\mathrm{M}(C(X'))$ *as an infinite matrix, and for each continuous function* $\phi : X \longrightarrow X'$, *define* $MC(\phi) : \mathrm{M}(C(X')) \longrightarrow \mathrm{M}(C(X))$ *by applying the functor* C *from Example 1.9.12 to each matrix entry. Then* MC *is a contravariant functor from the category of compact Hausdorff spaces to* $\mathfrak{MC}$.

Example 1.9.14 *In Example 1.9.13 the map* $MC(\phi)$ *maps idempotents to idempotents for every* ϕ *and thus we have a contravariant functor* Idem *from the category of compact Hausdorff spaces to the category of abelian monoids.*

In Examples 1.9.11 through 1.9.14, we constructed various contravariant functors F from the category of compact Hausdorff spaces to other categories. In such cases, given a continuous function $\phi : X \longrightarrow X'$, we will write $F(\phi)$ as ϕ^*.

Definition 1.9.15 *Let* $\mathfrak{A}$ *and* $\mathfrak{B}$ *be categories, and let* F *and* G *be covariant functors from* $\mathfrak{A}$ *to* $\mathfrak{B}$. *A natural transformation* $\mathbf{T}$ *from* F *to*

G is a function that assigns to each object A in $\mathfrak{A}$ a morphism $\mathbf{T}(A)$: $F(A) \longrightarrow G(A)$ in $\mathfrak{B}$ with the following property: for every object A' in $\mathfrak{A}$ and every morphism α in $\mathrm{Hom}(A, A')$, the diagram

$$
\begin{array}{ccc}
F(A) & \xrightarrow{\ \mathbf{T}(A)\ } & G(A) \\
{\scriptstyle F(\alpha)}\downarrow & & \downarrow{\scriptstyle G(\alpha)} \\
F(A') & \xrightarrow{\ \mathbf{T}(A')\ } & G(A')
\end{array}
$$

commutes.

Similarly, let $\widetilde{F}$ and $\widetilde{G}$ be contravariant functors from $\mathfrak{A}$ to $\mathfrak{B}$. A natural transformation $\widetilde{\mathbf{T}}$ from $\widetilde{F}$ to $\widetilde{G}$ is a function that assigns to each object A in $\mathfrak{A}$ a morphism $\widetilde{\mathbf{T}}(A) : \widetilde{F}(A) \longrightarrow \widetilde{G}(A)$ in $\mathfrak{B}$ with the property that for every object A' in $\mathfrak{A}$ and every morphism α in $\mathrm{Hom}(A, A')$, the diagram

$$
\begin{array}{ccc}
\widetilde{F}(A') & \xrightarrow{\ \widetilde{\mathbf{T}}(A')\ } & \widetilde{G}(A') \\
{\scriptstyle \widetilde{F}(\alpha)}\downarrow & & \downarrow{\scriptstyle \widetilde{G}(\alpha)} \\
\widetilde{F}(A) & \xrightarrow{\ \widetilde{\mathbf{T}}(A)\ } & \widetilde{G}(A)
\end{array}
$$

commutes.

If the categories $\mathfrak{A}$ and $\mathfrak{B}$ are clear from context, we usually do not explicitly mention them. Also, we often shorten the term "natural transformation" to "natural."

1.10 Notes

Almost all the results in this chapter are well known and/or "folk theorems"; many of the proofs are adapted from [5] or [16].

We have discussed only the most rudimentary ideas about homological algebra and category theory; for a more thorough introduction to these ideas, the reader should consult [10].

Exercises

1.1   Let X be a compact Hausdorff space and let n be a natural number. Idempotents E and F in $\mathrm{M}(n, C(X))$ are *algebraically equivalent* if there exist elements A and B in $\mathrm{M}(n, C(X))$ such that $\mathsf{AB} = \mathsf{E}$ and $\mathsf{BA} = \mathsf{F}$.

(a) Show that A and B can be chosen so that A = EA = AF = EAF and B = FB = BE = FBE.

(b) Prove that algebraic equivalence defines an equivalence relation $\sim_a$ on the idempotents in $M(n, C(X))$.

(c) Prove that $E \sim_s F$ implies that $E \sim_a F$.

(d) If E, F in $M(n, C(X))$ are algebraically equivalent, prove that $\text{diag}(E, 0_n)$ is similar to $\text{diag}(F, 0_n)$ in $M(2n, C(X))$.

(e) Conclude that algebraic equivalence, similarity, and homotopy coincide on $\text{Idem}(C(X))$.

1.2   Let X and Y be compact Hausdorff spaces, let V and V' be vector bundles over Y, and suppose that $\alpha : V \longrightarrow V'$ is a bundle homomorphism.

(a) Show that a continuous map $\phi : X \longrightarrow Y$ determines a bundle homomorphism $\phi^*\alpha : \phi^*V \longrightarrow \phi^*V'$.

(b) Verify that ϕ^* satisfies the following two properties:

   1. If $V = V'$ and α is the identity map on V, then $\phi^*\alpha$ is the identity map on ϕ^*V.
   2. If $\alpha' : V' \longrightarrow V''$ is a bundle homomorphism, then $\phi^*(\alpha'\alpha) = (\phi^*\alpha')(\phi^*\alpha)$.

1.3   Let X be a compact Hausdorff space, and define maps ϕ_0 and ϕ_1 from X to $X \times [0,1]$ by the formulas $\phi_0(x) = (x,0)$ and $\phi_1(x) = (x,1)$. We say that vector bundles V_0 and V_1 over X are *homotopic* if there exists a vector bundle W over $X \times [0,1]$ with the property that $\phi_0^*W \cong V_0$ and $\phi_1^*W \cong V_1$. Show that if V_0 and V_1 are homotopic, then $[V_0] = [V_1]$ in $\text{Vect}(X)$.

1.4   Suppose that V and W are vector bundles over a compact Hausdorff space X.

(a) Show that the collection $\text{Hom}(V, W)$ of bundle homomorphisms from V to W can be endowed with a topology that makes it into a vector bundle over X.

(b) Show that there is a one-to-one correspondence between vector bundle homomorphisms $\gamma : V \longrightarrow W$ and continuous functions $\Gamma : X \longrightarrow \text{Hom}(V, W)$.

1.5   Let V and W be vector bundles over a compact Hausdorff space X. Give an example to show that the kernel of ϕ need not be a subbundle of V and that the range of ϕ need not be a subbundle of W.

1.6 Suppose V is a vector bundle over a compact Hausdorff space X, and let W be a subbundle of V. For each x in X, form the quotient vector space $Q_x = V_x/W_x$, and let Q be the family $\{Q_x\}$ of vector spaces over X. Show that with the quotient topology, the family Q is a vector bundle.

1.7 Let A_N and A_S be the closed northern hemisphere and closed southern hemisphere of S^2, respectively, and let $Z = A_N \cap A_S$ be the equator.

    (a) Let $\widetilde{\phi} : S^1 \longrightarrow \mathrm{GL}(n, \mathbb{C})$ be a continuous functions. Show that $\widetilde{\phi}$ determines a clutching map $\phi : \Theta^n(A_N)|Z \longrightarrow \Theta^n(A_S)|Z$ and thus a vector bundle $\Theta^n(A_N) \cup_\phi \Theta^n(A_S)$ over S^2.

    (b) Prove that up to bundle isomorphism, every rank n vector bundle on S^2 can be constructed in this way.

    (c) If ϕ_0 and ϕ_1 are homotopic maps from S^1 to $\mathrm{GL}(n, \mathbb{C})$, show that the vector bundles $\Theta^n(A_N) \cup_{\phi_0} \Theta^n(A_S)$ and $\Theta^n(A_N) \cup_{\phi_1} \Theta^n(A_S)$ are isomorphic.

    (d) Generalize to higher-dimensional spheres.

1.8 Let V be a vector bundle over a compact Hausdorff space X, and suppose W is a subbundle of V. Choose two Hermitian metrics $\langle \, , \, \rangle_0$ and $\langle \, , \, \rangle_0$ on X and for $k = 0, 1$, let $W^{\perp,k}$ be the orthogonal complement of W with respect to $\langle \, , \, \rangle_k$ in V. Prove that $[W^{\perp,0}] = [W^{\perp,1}]$ in $\mathrm{Vect}(X)$. Hint: Use Exercise 1.3.

1.9 (For readers with some knowledge of differential topology.) Fix a natural number m, and let E be the idempotent defined in Example 1.4.4. Show that the vector bundle $\mathrm{Ran}(I - \mathsf{E})$ is isomorphic to the complexified tangent bundle of S^m.

1.10 An abelian monoid $\mathcal{A}$ has *cancellation* if $a + c = b + c$ implies that $a = b$ for all a, b, c in $\mathcal{A}$. Show that the following statements are equivalent:

    (a) The abelian monoid $\mathcal{A}$ has cancellation.

    (b) The equation $a_0 - a_1 = b_0 - b_1$ in $\mathcal{G}(\mathcal{A})$ implies $a_0 + b_1 = a_1 + b_0$ in $\mathcal{A}$.

    (c) The monoid homomorphism $j : \mathcal{A} \longrightarrow \mathcal{G}(\mathcal{A})$ in the statement of Theorem 1.6.7 is injective.

1.11 Verify that the sequence of groups in Example 1.8.3 is exact, but not split.

1.12 Fix a natural number m, and let E be the idempotent defined in Example 1.4.4. Show that $[\mathsf{E}] + [\Theta^1(S^m)] = [\Theta^{m+1}(S^m)]$ in $\mathrm{Idem}(S^m)$.

1.13   Let (V, π) be a vector bundle over a compact topological space X. A *section* of V is a continuous map $s : X \longrightarrow V$ such that $\pi(s(x)) = x$ for all x in X. Show that V is a trivial bundle of rank n if and only if there exist nonvanishing sections $s_1, s_2, \ldots, s_n$ of V such that for each x in X, the set $\{s_1(x), s_2(x), \ldots, s_k(x)\}$ is a vector space basis for V_x.

1.14   Let V be a vector bundle over a compact Hausdorff space X.

(a) Suppose that A is a closed subspace of X. Show that any section of $V|A$ extends to a section of V. (Hint: first prove the result for trivial bundles, and then use a partition of unity argument.)

(b) Let x_1 and x_2 be distinct points of X, and suppose v_1 and v_2 are points of V such that $\pi(v_1) = x_1$ and $\pi(v_2) = x_2$. Prove that there exists a section s of V such that $s(x_1) = v_1$ and $s(x_2) = v_2$.

2

K-theory

In this chapter we define the various K-theory groups associated to a topological space and study how they are related.

2.1 Definition of $\mathrm{K}^0(X)$

Definition 2.1.1 *Let X be compact Hausdorff. The Grothendieck completion of* $\mathrm{Vect}(X)$ *is denoted* $\mathrm{K}^0(X)$.

Thanks to Theorem 1.7.14, we may alternately define $\mathrm{K}^0(X)$ as the Grothendieck completion of $\mathrm{Idem}(C(X))$.

Example 2.1.2 *Let X be a single point. Then a vector bundle over X is just a vector space, and these are classified by rank. Thus* $\mathrm{Vect}(X) \cong \mathbb{Z}^+$ *and so* $\mathrm{K}^0(X) \cong \mathbb{Z}$.

Example 2.1.3 *Let X be the disjoint union of compact Hausdorff spaces X_1, X_2, ..., X_k. A vector bundle on X is a choice of a vector bundle on each X_1, X_2, ..., X_k, and the same is true for isomorphism classes of vector bundles on X. Therefore*

$$\mathrm{Vect}(X) \cong \mathrm{Vect}(X_1) \oplus \mathrm{Vect}(X_2) \oplus \cdots \oplus \mathrm{Vect}(X_k).$$

By taking the Grothendieck completion, we obtain an isomorphism

$$\mathrm{K}^0(X) \cong \mathrm{K}^0(X_1) \oplus \mathrm{K}^0(X_2) \oplus \cdots \oplus \mathrm{K}^0(X_k).$$

In particular, if X consists of k distinct points in the discrete topology, then $\mathrm{K}^0(X) \cong \mathbb{Z}^k$.

We would certainly like to be able to compute $K^0(X)$ for a topological space more complicated than a finite set of points! To do this, we have to develop some machinery; this will occupy us for the remainder of the chapter.

Proposition 2.1.4 *Let X be a compact Hausdorff space.*

(i) *Every element of $K^0(X)$ can be written in the form $[E] - [I_k]$, where E is an idempotent in $M(n, C(X))$ and $n \geq k$.*

(ii) *Every element of $K^0(X)$ can be written in the form $[V] - [\Theta^k(X)]$ for some vector bundle V whose rank on each connected component of X is greater than or equal to k.*

Proof Let $[E_0] - [E_1]$ be an element of $K^0(X)$, with E_1 in $M(k, C(X))$. Lemma 1.7.3 implies that

$$[E_0] - [E_1] = [E_0] - ([E_1] + [I_k - E_1]) + [I_k - E_1]$$
$$= [E_0] - [I_k] + [I_k - E_1]$$
$$= [\mathrm{diag}(E_0, I_k - E_1)] - [I_k],$$

and so we can take $E = \mathrm{diag}(E_0, I_k - E_1)$. This proves (i); the proof of (ii) then follows easily from Proposition 1.7.6.   □

Unless stated otherwise, we will write elements of $K^0(X)$ in the form described in Proposition 2.1.4.

Proposition 2.1.5 *Let X be a compact Hausdorff space.*

(i) *Suppose E is an idempotent in $M(n, C(X))$. Then $[E] - [I_k] = 0$ in $K^0(X)$ if and only if there exists a natural number m such that $\mathrm{diag}(E, I_m, 0_m) \sim_s \mathrm{diag}(I_{k+m}, 0_{m+n-k})$ in $M(n + 2m, C(X))$.*

(ii) *Suppose V is a vector bundle over X. Then $[V] - [\Theta^k(X)] = 0$ in $K^0(X)$ if and only if there exists a natural number m such that $V \oplus \Theta^m(X)$ is isomorphic to $\Theta^{m+k}(X)$.*

Proof We shall only prove (i); the proof of (ii) is similar. Suppose that $\mathrm{diag}(E, I_m, 0_m) \sim_s \mathrm{diag}(I_{k+m}, 0_{m+n-k})$. Then

$$0 = [\mathrm{diag}(E, I_m, 0_m)] - [\mathrm{diag}(I_{k+m}, 0_{m+n-k})]$$
$$= ([E] + [I_m] + [0_m]) - ([I_k] + [I_m] + [0_{m+n-k}]) = [E] - [I_k].$$

Conversely, suppose $[E] - [I_k] = 0$. Then there exists a natural number m and an idempotent F in $M(m, C(X))$ such that $[\mathrm{diag}(\mathsf{E}, \mathsf{F})] = [\mathrm{diag}(I_k, \mathsf{F})]$ in $\mathrm{Idem}(C(X))$. Therefore

$$[\mathrm{diag}(\mathsf{E}, \mathsf{F}, I_m - \mathsf{F})] = [\mathrm{diag}(I_k, \mathsf{F}, I_m - \mathsf{F})]$$

in $\mathrm{Idem}(C(X))$, and thus $[\mathrm{diag}(\mathsf{E}, I_m)] = [\mathrm{diag}(I_k, I_m)] = [I_{k+m}]$ by Lemma 1.7.3. From the definition of $M(C(X))$ and $\mathrm{Idem}(C(X))$, we know that $\mathrm{diag}(\mathsf{E}, I_m, 0_l)$ and $\mathrm{diag}(I_{k+m}, 0_j)$ are similar for some natural numbers l and j, and by comparing matrix sizes, we see that $j = l+n-k$. By either increasing l, or increasing m if necessary and making the appropriate adjustment to F, we may take $l = m$ and $j = m + n - k$, which proves the result. $\qquad\qquad\square$

Proposition 2.1.6 *The assignment $X \mapsto K^0(X)$ determines a contravariant functor K^0 from the category of compact Hausdorff spaces to the category of abelian groups.*

Proof Example 1.9.11 gives us a contravariant functor Vect from the category of topological spaces to the category of abelian monoids. If we restrict Vect to the category of compact Hausdorff spaces and compose with the Grothendieck completion functor from Example 1.9.10, we obtain K^0. $\qquad\qquad\square$

In less fancy language, Proposition 2.1.6 states the following: suppose X, Y, and Z are compact Hausdorff spaces.

- If $\phi : X \longrightarrow Y$ is continuous, then pullback of vector bundles induces a group homomorphism $\phi^* : K^0(Y) \longrightarrow K^0(X)$.
- Let id_X denote the identity map on X. Then $(\mathrm{id}_X)^*$ is the identity map on $K^0(X)$.
- If $\phi : X \longrightarrow Y$ and $\psi : Y \longrightarrow Z$ are continuous, then $(\psi \circ \phi)^* = \phi^* \circ \psi^*$.

Recall that topological spaces X and Y are *homotopy equivalent* if there exist continuous functions $\phi : X \longrightarrow Y$ and $\psi : Y \longrightarrow X$ such that $\psi\phi$ and $\phi\psi$ are homotopic to the identity maps id_X and id_Y on X and Y respectively. Also, a topological space X is *contractible* if the identity map on X is homotopic to a constant map.

Proposition 2.1.7 *Let X and Y be compact Hausdorff spaces.*

(i) *Suppose ϕ_0 and ϕ_1 are homotopic functions from X to Y. Then ϕ_0^* and ϕ_1^* from $K^0(Y)$ to $K^0(X)$ are equal homomorphisms.*

(ii) *If X and Y are homotopy equivalent, then $K^0(X) \cong K^0(Y)$.*

Proof Take an element $[\mathsf{E}] - [I_k]$ of $\mathrm{K}^0(Y)$. Obviously $\phi_0^*[I_k] = \phi_1^*[I_k] = [I_k]$ in $\mathrm{K}^0(X)$. Let $\{\phi_t\}$ be a homotopy from ϕ_0 to ϕ_1. Then $\{\phi_t^*\mathsf{E}\}$ is a homotopy of idempotents, so $[\phi_0^*\mathsf{E}] = [\phi_1^*\mathsf{E}]$. This proves (i); we obtain (ii) from (i) and the definition of homotopy equivalence.   □

Corollary 2.1.8 *Let X be a contractible compact Hausdorff space. Then $\mathrm{K}^0(X) \cong \mathbb{Z}$.*

Proof Choose a point x_0 in X so that there is a homotopy from the identity map on X to the constant map on X whose range is $\{x_0\}$. Then Example 2.1.2 and Proposition 2.1.7 imply that $\mathrm{K}^0(X) \cong \mathrm{K}^0(\{x_0\}) \cong \mathbb{Z}$.   □

Example 2.1.9 *Define $\phi_t(x) = (1-t)x$ for each $0 \leq t \leq 1$ and x in X. Then $\{\phi_t\}$ is a homotopy from the identity map on $[0,1]$ to the constant function 0, and thus $\mathrm{K}^0([0,1]) \cong \mathbb{Z}$.*

2.2 Relative K-theory

An important technique in understanding the K-theory of a compact Hausdorff space is to relate it to the K-theory of its closed subspaces. In this section, we make this notion precise.

Definition 2.2.1 *A compact pair (X, A) is a compact Hausdorff space X and a closed subspace A of X. A morphism $\phi : (X, A) \longrightarrow (Y, B)$ of compact pairs is a continuous function $\phi : X \longrightarrow Y$ such that $\phi(A) \subseteq B$.*

We will assume that A is nonempty unless explicitly specified. When A consists of a single point x_0, we will usually simplify notation by writing (X, x_0) instead of the more technically correct $(X, \{x_0\})$.

Definition 2.2.2 *Let (X_1, A_1) and (X_2, A_2) be copies of a compact pair (X, A), and form the disjoint union $X_1 \coprod X_2$. Identify each point of A_1 with its corresponding point in A_2. The resulting set $\mathcal{D}(X, A)$, endowed with the quotient topology, is called the* double *of X along A.*

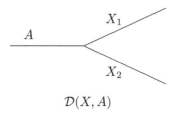

$$\mathcal{D}(X, A)$$

Proposition 2.2.3 *The assignment* $(X, A) \mapsto \mathcal{D}(X, A)$ *determines a covariant functor from compact pairs to compact Hausdorff spaces.*

Proof The topological space $\mathcal{D}(X, A)$ is compact because it is the image of the compact space $X_1 \coprod X_2$, and an easy argument by cases shows that $\mathcal{D}(X, A)$ is Hausdorff. Suppose $\phi : (X, A) \longrightarrow (Y, B)$ is a morphism of compact pairs. Then ϕ defines in the obvious way a continuous map $\widehat{\phi} : X_1 \coprod X_2 \longrightarrow Y_1 \coprod Y_2$. Because $\phi(A)$ is contained in B, the map $\widehat{\phi}$ uniquely determines a continuous function $\mathcal{D}\phi : \mathcal{D}(X, A) \longrightarrow \mathcal{D}(Y, B)$ that satisfies the axioms for a covariant functor. $\qquad\square$

We will write matrices over $\mathcal{D}(X, A)$ as pairs (B_1, B_2), where B_1 and B_2 are matrices over X such that $B_1(a) = B_2(a)$ for all a in A. Given an idempotent (E_1, E_2) over $\mathcal{D}(X, A)$, we will drop the parentheses and let $[E_1, E_2]$ represent the corresponding element of $K^0(\mathcal{D}(X, A))$. We record here a simple but useful fact about elements of $K^0(\mathcal{D}(X, A))$.

Lemma 2.2.4 *Let* (X, A) *be a compact pair and let* E_1 *and* E_2 *be idempotents in* $M(n, C(X))$. *Suppose that* (E_1, SE_2S^{-1}) *is an idempotent in* $M(n, C(\mathcal{D}(X, A)))$ *for some* S *in* $GL(n, C(X))$. *Then* $(S^{-1}E_1S, E_2)$ *is an idempotent in* $M(n, C(\mathcal{D}(X, A)))$ *and* $[E_1, SE_2S^{-1}] = [S^{-1}E_1S, E_2]$ *in* $K^0(\mathcal{D}(X, A))$.

Proof Obviously (S, S) is an element in $GL(n, C(\mathcal{D}(X, A)))$, and

$$(S, S)^{-1}(E_1, SE_2S)(S, S) = (S^{-1}E_1S, E_2).$$

$\qquad\square$

Definition 2.2.5 *Let* (X, A) *be a compact pair. For* $i = 1, 2$, *we define maps* $\mu_i : X \longrightarrow \mathcal{D}(X, A)$ *by composing the inclusion of* X *into* X_i *with the quotient map from* $X_1 \coprod X_2$ *onto* $\mathcal{D}(X, A)$.

Definition 2.2.6 *Let (X, A) be a compact pair. The* relative K^0*-group* $K^0(X, A)$ *is the kernel of the homomorphism*

$$\mu_2^* : K^0(\mathcal{D}(X, A)) \longrightarrow K^0(X).$$

Proposition 2.2.7 *Relative K^0 defines a contravariant functor from the category of compact pairs to the category of abelian groups.*

Proof Let $\phi : (X, A) \longrightarrow (Y, B)$ be a morphism of compact pairs. By Proposition 2.2.3 and the definition of relative K^0, the diagram

$$
\begin{array}{ccccccc}
0 & \longrightarrow & K^0(Y, B) & \longrightarrow & K^0(\mathcal{D}(Y, B)) & \xrightarrow{\mu_2^*} & K^0(Y) \\
& & & & \downarrow{\scriptstyle (\mathcal{D}\phi)^*} & & \downarrow{\scriptstyle \phi^*} \\
0 & \longrightarrow & K^0(X, A) & \longrightarrow & K^0(\mathcal{D}(X, A)) & \xrightarrow{\mu_2^*} & K^0(X)
\end{array}
$$

commutes and has exact rows. Proposition 1.8.9 implies that $(\mathcal{D}\phi)^*$ restricts to a homomorphism from $K^0(Y, B)$ to $K^0(X, A)$. Furthermore, the commutativity of the diagram and Proposition 2.2.3 yield that relative K^0 is a contravariant functor. □

Proposition 2.2.8 *Let (X, A) be a compact pair. Then every element of $K^0(X, A)$ can be written in the form $[\mathsf{E}, \mathrm{diag}(I_k, 0_k)] - [I_k, I_k]$ for some natural number k.*

Proof By Proposition 2.1.4, every element of $K^0(\mathcal{D}(X, A))$ has the form $[\mathsf{E}_1, \mathsf{E}_2] - [I_j, I_j]$, where $(\mathsf{E}_1, \mathsf{E}_2)$ is an idempotent in $M(n, C(\mathcal{D}(X, A)))$ and $n \geq j$. If $[\mathsf{E}_1, \mathsf{E}_2] - [I_j, I_j]$ is in $K^0(X, A)$, then $[\mathsf{E}_2] - [I_j] = 0$ in $K^0(X)$. Next, Proposition 2.1.5 implies that for some natural number m, there is a matrix S in $GL(n+2m, C(X))$ such that $\mathrm{diag}(\mathsf{E}_2, I_m, 0_m) = \mathsf{S}\,\mathrm{diag}(I_{j+m}, 0_{m+n-j})\mathsf{S}^{-1}$. Lemma 2.2.4 gives us

$$
\begin{aligned}
& [\mathsf{E}_1, \mathsf{E}_2] - [I_j, I_j] \\
& = [\mathrm{diag}(\mathsf{E}_1, I_m, 0_m), \mathrm{diag}(\mathsf{E}_2, I_m, 0_m)] - [I_{j+m}, I_{j+m}] \\
& = [\mathrm{diag}(\mathsf{E}_1, I_m, 0_m), \mathsf{S}\,\mathrm{diag}(I_{j+m}, 0_{m+n-j})\mathsf{S}^{-1}] - [I_{j+m}, I_{j+m}] \\
& = [\mathsf{S}^{-1}\,\mathrm{diag}(\mathsf{E}_1, I_m, 0_m)\mathsf{S}, \mathrm{diag}(I_{j+m}, 0_{m+n-j})] - [I_{j+m}, I_{j+m}].
\end{aligned}
$$

Let $\widehat{\mathsf{E}} = \mathsf{S}^{-1}\,\mathrm{diag}(\mathsf{E}_1, I_m, 0_m)\mathsf{S}$. Then for every natural number l, Lemmas 2.2.4 and 1.7.1 state that for some U in $GL(n + 2m + l, \mathbb{C})$, we have

$$[\widehat{\mathsf{E}}, \operatorname{diag}(I_{j+m}, 0_{m+n-j})] - [I_{j+m}, I_{j+m}]$$
$$= [\operatorname{diag}(\widehat{\mathsf{E}}, I_l), \operatorname{diag}(I_{j+m}, 0_{m+n-j}, I_l)] - [I_{j+m+l}, I_{j+m+l}]$$
$$= [\operatorname{diag}(\widehat{\mathsf{E}}, I_l), \mathsf{U} \operatorname{diag}(I_{j+m+l}, 0_{m+n-j}) \mathsf{U}^{-1}] - [I_{j+m+l}, I_{j+m+l}]$$
$$= [\mathsf{U}^{-1} \operatorname{diag}(\widehat{\mathsf{E}}, I_l) \mathsf{U}, \operatorname{diag}(I_{j+m+l}, 0_{m+n-j})] - [I_{j+m+l}, I_{j+m+l}].$$

Choose l large enough so that $2j - n + l$ is positive. Then

$$[\mathsf{U}^{-1} \operatorname{diag}(\widehat{\mathsf{E}}, I_l) \mathsf{U}, \operatorname{diag}(I_{j+m+l}, 0_{m+n-j})] - [I_{j+m+l}, I_{j+m+l}]$$
$$= [\operatorname{diag}(\mathsf{U}^{-1} \operatorname{diag}(\widehat{\mathsf{E}}, I_l) \mathsf{U}, 0_{2j-n+l}), \operatorname{diag}(I_{j+m+l}, 0_{m+n-j}, 0_{2j-n-l})]$$
$$- [I_{j+m+l}, I_{j+m+l}]$$
$$= [\operatorname{diag}(\mathsf{U}^{-1} \operatorname{diag}(\widehat{\mathsf{E}}, I_l) \mathsf{U}, 0_{2j-n+l}), \operatorname{diag}(I_{j+m+l}, 0_{j+m+l})]$$
$$- [I_{j+m+l}, I_{j+m+l}].$$

The proposition follows by setting $k = j + m + l$ and

$$\mathsf{E} = \operatorname{diag}(\mathsf{U}^{-1} \operatorname{diag}(\widehat{\mathsf{E}}, I_l) \mathsf{U}, 0_{2j-n+l}).$$

$\square$

Theorem 2.2.9 *Let (X, A) be a compact pair and let $j : A \longrightarrow X$ be the inclusion map. Then the sequence*

$$K^0(X, A) \xrightarrow{\mu_1^*} K^0(X) \xrightarrow{j^*} K^0(A)$$

is exact.

Proof We are slightly abusing notation here; the map from $K^0(X, A)$ to $K^0(X)$ is obtained by restricting the domain of $\mu_1^* : K^0(\mathcal{D}(X, A)) \longrightarrow K^0(X)$ to $K^0(X, A)$, but no harm should be caused by also calling the restricted map μ_1^*.

Take $[\mathsf{E}, \operatorname{diag}(I_k, 0_k)] - [I_k, I_k]$ in $K^0(X, A)$. Then

$$j^* \mu_1^* ([\mathsf{E}, \operatorname{diag}(I_k, 0_k)] - [I_k, I_k]) = j^* ([\mathsf{E}] - [I_k])$$
$$= [\mathsf{E}|A] - [I_k]$$
$$= [\operatorname{diag}(I_k, 0_k)] - [I_k]$$
$$= 0.$$

Thus the image of μ_1^* is contained in the kernel of j^*. To show the reverse containment, take $[\mathsf{F}] - [I_l]$ in $K^0(X)$ with F in $M(n, C(X))$ for some natural number n, and suppose that $j^*([\mathsf{F}]-[I_l]) = [\mathsf{F}|A]-[I_l] = 0$ in $\mathrm{K}^0(A)$. Proposition 2.1.5(i) states for some natural number m, there is an element S in $\mathrm{GL}(n+2m,C(A))$ such that

$$\mathsf{S}\,\mathrm{diag}(\mathsf{F}|A,I_m,0_m)\mathsf{S}^{-1} = \mathrm{diag}(I_{l+m},0_{m+n-l}).$$

By Corollary 1.3.17, we may choose T in $\mathrm{GL}(2(n+2m),C(X))$ with the property that $\mathsf{T}|A = \mathrm{diag}(\mathsf{S},\mathsf{S}^{-1})$. Define

$$\widehat{\mathsf{F}} = \mathsf{T}\,\mathrm{diag}(\mathsf{F},I_m,0_m,0_{n+2m})\mathsf{T}^{-1}.$$

Then

$$\widetilde{\mathsf{F}} = \left[\widehat{\mathsf{F}},\mathrm{diag}(I_{l+m},0_{m+n-l},0_{n+2m})\right] - [I_{l+m},I_{l+m}]$$

is an element of $\mathrm{K}^0(\mathcal{D}(X,A))$. Now, $\mu_2^*\widetilde{\mathsf{F}} = 0$, and thus $\widetilde{\mathsf{F}}$ is in $\mathrm{K}^0(X,A)$. We complete the proof by computing

$$
\begin{aligned}
\mu_1^*\widetilde{\mathsf{F}} &= [\widehat{\mathsf{F}}] - [I_{l+m}] \\
&= [T\,\mathrm{diag}(\mathsf{F},I_m,0_m,0_{n+2m})T^{-1}] - [I_{l+m}] \\
&= [\mathrm{diag}(\mathsf{F},I_m,0_m,0_{n+2m})] - [I_{l+m}] \\
&= [\mathsf{F}] + [I_m] - ([I_l]+[I_m]) \\
&= [\mathsf{F}] - [I_l].
\end{aligned}
$$

$\square$

By definition, Theorem 2.2.9 states that K^0 is a *half-exact functor* (although the author thinks this property is more appropriately called *third*-exactness). In general, μ_1^* is not injective and j^* is not surjective, so it is not always possible to put 0s on the ends of the sequence in the statement of Theorem 2.2.9. However, there is an important special case in which we can do this.

Theorem 2.2.10 *Let (X,A) be a compact pair, let $j : A \longrightarrow X$ be the inclusion map, and suppose there exists a continuous function $\psi : X \longrightarrow A$ such that ψj is the identity map on A. Then there exists a split exact sequence*

$$0 \longrightarrow \mathrm{K}^0(X,A) \xrightarrow{\ \mu_1^*\ } \mathrm{K}^0(X) \underset{\psi^*}{\overset{j^*}{\rightleftarrows}} \mathrm{K}^0(A) \longrightarrow 0,$$

whence $\mathrm{K}^0(X)$ is isomorphic to $\mathrm{K}^0(A) \oplus \mathrm{K}^0(X,A)$.

Proof We prove that the sequence in the statement of the theorem is split exact; the isomorphism $K^0(X) \cong K^0(A) \oplus K^0(X, A)$ is a consequence of Proposition 1.8.7.

We already know from Theorem 2.2.9 that the sequence is exact at $K^0(X)$. Because ψj is the identity on A, the homomorphism $(\psi j)^* = j^* \psi^*$ is the identity on $K^0(A)$, which implies that j^* is surjective and hence we have exactness at $K^0(A)$.

To show that the sequence is exact at $K^0(X, A)$, take $[E, \mathrm{diag}(I_k, 0_k)] - [I_k, I_k]$ in $K^0(X, A)$, and suppose that

$$\mu_1^*\big([E, \mathrm{diag}(I_k, 0_k)] - [I_k, I_k]\big) = [E] - [I_k] = 0$$

in $K^0(X)$. By Proposition 2.1.5(i), there exists a natural number m and and an element S in $GL(2k + 2m, C(X))$ such that $\mathrm{diag}(I_{k+m}, 0_{k+m}) = S\,\mathrm{diag}(E, I_m, 0_m)S^{-1}$. We also have

$$
\begin{aligned}
\mathrm{diag}(I_{k+m}, 0_{k+m}) &= \psi^* j^* \,\mathrm{diag}(I_{k+m}, 0_{k+m}) \\
&= \psi^* j^* \big(S\,\mathrm{diag}(E, I_m, 0_m)S^{-1}\big) \\
&= (\psi^* j^* S)\,\mathrm{diag}(\psi^* j^* E, I_m, 0_m)(\psi^* j^* S)^{-1}.
\end{aligned}
$$

Note that $(\psi^* j^* S)|A = j^* \psi^* j^* S = j^* S = S|A$, so $(S, \psi^* j^* S)$ is an element of $GL(2k + 2m, C(\mathcal{D}(X, A)))$. The same reasoning shows that

$$\big(\mathrm{diag}(E, I_m, 0_m), \psi^* j^*(\mathrm{diag}(E, I_m, 0_m))\big) = $$
$$\big(\mathrm{diag}(E, I_m, 0_m), \mathrm{diag}(\psi^* j^* E, I_m, 0_m)\big)$$

is an element of $M(2k + 2m, C(\mathcal{D}(X, A)))$. Thus

$$
\begin{aligned}
(S, \psi^* j^* S)&\big(\mathrm{diag}(E, I_m, 0_m), \mathrm{diag}(\psi^* j^* E, I_m, 0_m)\big)(S, \psi^* j^* S)^{-1} \\
&= \big(S\,\mathrm{diag}(E, I_m, 0_m)S^{-1}, (\psi^* j^* S)\,\mathrm{diag}(\psi^* j^* E, I_m, 0_m)(\psi^* j^* S)^{-1}\big) \\
&= \big(S\,\mathrm{diag}(E, I_m, 0_m)S^{-1}, \psi^* j^*(S\,\mathrm{diag}(E, I_m, 0_m)S^{-1})\big) \\
&= \big(\mathrm{diag}(I_{k+m}, 0_{k+m}), \mathrm{diag}(I_{k+m}, 0_{k+m})\big),
\end{aligned}
$$

and therefore $[E, \psi^* j^* E] = [I_k, I_k]$ in $K^0(\mathcal{D}(X, A))$. But

$$\psi^* j^* E = \psi^*(E|A) = \psi^*\,\mathrm{diag}(I_k, 0_k) = \mathrm{diag}(I_k, 0_k),$$

whence

$$[E, \mathrm{diag}(I_k, 0_k)] = [E, \psi^* j^* E] = [I_k, I_k]$$

in $K^0(\mathcal{D}(X, A))$. Therefore $[E, \mathrm{diag}(I_k, 0_k)] - [I_k, I_k] = 0$ in $K^0(X, A)$, and so μ_1^* is injective. $\qquad\square$

Corollary 2.2.11 *Let X be a compact Hausdorff space and take x_0 in X. Then*

$$\mathrm{K}^0(X) \cong \mathrm{K}^0(X, x_0) \oplus \mathrm{K}^0(x_0) \cong \mathrm{K}^0(X, x_0) \oplus \mathbb{Z}.$$

Proof The sequence

$$0 \longrightarrow \mathrm{K}^0(X, x_0) \xrightarrow{\ \mu_1^*\ } \mathrm{K}^0(X) \underset{\psi^*}{\overset{j^*}{\rightleftarrows}} \mathrm{K}^0(\{x_0\}) \longrightarrow 0$$

is split exact, and $\mathrm{K}^0(\{x_0\}) \cong \mathbb{Z}$ by Example 2.1.2. $\square$

An important property of $\mathrm{K}^0(X, A)$ is that it only depends, up to isomorphism, on the complement of A in X. We begin with a definition.

Definition 2.2.12 *For any compact pair (X, A), we define $(X\backslash A)^+$ to be the topological space obtained by identifying all the points of A to a single point ∞ and equipping the resulting set with the quotient topology. We let q denote the quotient map from X to $(X\backslash A)^+$.*

The reader can check that $(X\backslash A)^+$ is a compact Hausdorff space. As we shall see in the next section, an alternate description of $(X\backslash A)^+$ is as the *one-point compactification* of $X\backslash A$.

Lemma 2.2.13 *Let (X, A) be a compact pair and let n be a natural number. The homomorphism*

$$q^* : \mathrm{M}(n, C((X\backslash A)^+)) \longrightarrow \mathrm{M}(n, C(X))$$

is injective, and its image is precisely the subset of matrices that are constant on A.

Proof The map q^* is applied entrywise to elements of $\mathrm{M}(n, C((X\backslash A)^+))$, so it suffices to prove the lemma for the case $n = 1$. If f is in $C((X\backslash A)^+)$ and $q^* f = 0$, then f itself is the zero function, whence q^* is injective. The definition of q implies that the image of q^* is contained in the set of functions on X that are constant on A. To see that q^* maps onto this set, suppose we have a function g in $C(X)$ and a complex number λ such that $g(a) = \lambda$ for all a in A. Define $\widehat{g} : (X\backslash A)^+ \longrightarrow \mathbb{C}$ as

$$\widehat{g}(z) = \begin{cases} \lambda & \text{if } z = \infty \\ g(z) & \text{if } z \in X\backslash A. \end{cases}$$

Then $\widehat{g}$ is continuous, and $q^* \widehat{g} = g$. $\square$

Theorem 2.2.14 *For every compact pair* (X, A), *the homomorphism*

$$q^* : K^0((X \backslash A)^+, \infty) \longrightarrow K^0(X, A)$$

is an isomorphism. Furthermore, this isomorphism is natural in the following sense: suppose that $\phi : (X, A) \longrightarrow (Y, B)$ *is a morphism of compact pairs and define* $\widetilde{\phi} : (X \backslash A)^+ \longrightarrow (Y \backslash B)^+$ *by the formula*

$$\widetilde{\phi}(z) = \begin{cases} z & \text{if } z \in X \backslash A \\ \infty & \text{if } z = \infty. \end{cases}$$

Then the diagram

$$
\begin{array}{ccc}
K^0((Y \backslash B)^+, \infty) & \xrightarrow{\ \widetilde{\phi}^* \ } & K^0((X \backslash A)^+, \infty) \\
\Big\downarrow{\scriptstyle q^*} & & \Big\downarrow{\scriptstyle q*} \\
K^0(Y, B) & \xrightarrow{\ \phi^* \ } & K^0(X, A)
\end{array}
$$

commutes.

Proof To show surjectivity, take $[\mathsf{E}, \operatorname{diag}(I_k, 0_k)] - [I_k, I_k]$ in $K^0(X, A)$. Then $\mathsf{E}(a) = \operatorname{diag}(I_k, 0_k)$ for all a in A. By Lemma 2.2.13, we have $\mathsf{E} = q^* \widetilde{\mathsf{E}}$ for some idempotent $\widetilde{\mathsf{E}}$ in $M(n, C((X \backslash A)^+))$, and so

$$q^* \big([\widetilde{\mathsf{E}}, \operatorname{diag}(I_k, 0_k)] - [I_k, I_k]\big) = [q^* \widetilde{\mathsf{E}}, \operatorname{diag}(I_k, 0_k)] - [I_k, I_k]$$
$$= [\mathsf{E}, \operatorname{diag}(I_k, 0_k)] - [I_k, I_k].$$

To show injectivity, take $[\mathsf{F}, \operatorname{diag}(I_k, 0_k)] - [I_k, I_k]$ in $K^0((X \backslash A)^+, \infty)$, and suppose that

$$q^* \big([\mathsf{F}, \operatorname{diag}(I_k, 0_k)] - [I_k, I_k]\big) = [q^* \mathsf{F}, \operatorname{diag}(I_k, 0_k)] - [I_k, I_k] = 0$$

in $K^0(X, A)$. Then for some natural number m, there is an element $(\mathsf{S}_1, \mathsf{S}_2)$ in $GL(n + 2m, C(\mathcal{D}(X, A)))$ such that

$$\mathsf{S}_1 \operatorname{diag}(q^* \mathsf{F}, I_m, 0_m) \mathsf{S}_1^{-1} = \operatorname{diag}(I_{k+m}, 0_{k+m})$$
$$= \mathsf{S}_2 \operatorname{diag}(I_k, 0_k, I_m, 0_m) \mathsf{S}_2^{-1}.$$

Thus

$$(\mathsf{S}_2^{-1}\mathsf{S}_1) \operatorname{diag}(q^* \mathsf{F}, I_m, 0_m)(\mathsf{S}_2^{-1}\mathsf{S}_1)^{-1} = \operatorname{diag}(I_k, 0_k, I_m, 0_m).$$

Because $\mathsf{S}_2^{-1}(a)\mathsf{S}_1(a) = I_{n+2m}$ for all a in A, Lemma 2.2.13 states that there exists a unique element T in $GL(n + 2m, C((X \backslash A)^+))$ such that

$q^*\mathsf{T} = \mathsf{S}_2^{-1}\mathsf{S}_1$. The injectivity of q^* implies

$$\mathsf{T}\,\mathrm{diag}(\mathsf{F}, I_m, 0_m)\mathsf{T}^{-1} = \mathrm{diag}(I_k, 0_k, I_m, 0_m),$$

and because (T, I_{n+2m}) is in $\mathrm{GL}(n+2m, C(\mathcal{D}((X\backslash A)^+, \infty)))$, we have

$$(\mathsf{T}, I_{n+2m})(\mathrm{diag}(\mathsf{F}, I_m, 0_m), \mathrm{diag}(I_k, 0_k, I_m, 0_m))(\mathsf{T}, I_{n+2m})^{-1}$$
$$= (\mathrm{diag}(I_k, 0_k, I_m, 0_m), \mathrm{diag}(I_k, 0_k, I_m, 0_m)).$$

Therefore

$$[\mathsf{F}, \mathrm{diag}(I_k, 0_k)] + [I_m, I_m]$$
$$= [\mathrm{diag}(\mathsf{F}, I_m, 0_m), \mathrm{diag}(I_k, 0_k, I_m, 0_m)]$$
$$= [\mathrm{diag}(I_k, 0_k, I_m, 0_m), \mathrm{diag}(I_k, 0_k, I_m, 0_m)]$$
$$= [\mathrm{diag}(I_{k+m}, 0_{k+m}), \mathrm{diag}(I_{k+m}, 0_{k+m})]$$
$$= [I_k, I_k] + [I_m, I_m]$$

by Lemma 1.7.1, and so $[\mathsf{F}, \mathrm{diag}(I_k, 0_k)] - [I_k, I_k] = 0$ in $\mathrm{K}^0((X\backslash A)^+, \infty)$. Finally, routine computations confirm that the diagram in the statement of the theorem commutes. $\square$

Corollary 2.2.15 (Excision for K^0) *Let (X, A) be a compact pair and let U be an open subspace of A. Then there exists a natural isomorphism*

$$\mathrm{K}^0(X\backslash U, A\backslash U) \cong \mathrm{K}^0(X, A).$$

Proof Let

$$q_1 : (X, A) \longrightarrow ((X\backslash A)^+, \infty)$$

and

$$q_2 : (X\backslash U, A\backslash U) \longrightarrow (((X\backslash U)\backslash(A\backslash U))^+, \infty)$$

be the quotient maps described in Definition 2.2.12. By Theorem 2.2.14, the maps q_1^* and q_2^* are isomorphisms. Then because $(X\backslash U)\backslash(A\backslash U) = X\backslash A$, the composition $q_1^*(q_2^*)^{-1}$ is well defined and is a natural isomorphism. $\square$

2.3 Invertibles and K^{-1}

In this section we introduce a second *K*-theory group, this one constructed from invertible elements.

Definition 2.3.1 *Let X be a compact Hausdorff space and define*

$$\widehat{GL}(C(X)) = \bigcup_{n \in \mathbb{N}} \text{GL}(n, C(X)).$$

Define an equivalence relation $\sim$ *on* $\widehat{GL}(C(X))$ *by decreeing that* $\mathsf{S} \sim$ diag$(\mathsf{S}, 1)$ *for all natural numbers n and matrices* S *in* GL$(n, C(X))$. *We define* GL$(C(X))$ *to be the set of equivalence classes of* $\sim$. *Set*

$$\widehat{GL}(C(X))_{\mathbf{0}} = \bigcup_{n \in \mathbb{N}} \text{GL}(n, C(X))_{\mathbf{0}},$$

and define GL$(C(X))_{\mathbf{0}}$ *to be the set of equivalence classes of* $\sim$ *restricted to* $\widehat{GL}(C(X))_{\mathbf{0}}$.

To simplify notation, we will blur the distinction between elements of GL$(n, C(X))$ and GL$(n, C(X))_{\mathbf{0}}$ and their respective images in the sets GL$(C(X))$ and GL$(C(X))_{\mathbf{0}}$.

Proposition 2.3.2 *Suppose that X is a compact Hausdorff space. Then* GL$(C(X))$ *is a group under matrix multiplication, and* GL$(C(X))_{\mathbf{0}}$ *is a normal subgroup of* GL$(C(X))$.

Proof Take elements S and T in GL$(C(X))$ and choose a natural number n large enough so that S and T can be represented by matrices in GL$(n, C(X))$. Then diag$(S, 1)$ diag$(T, 1) = $ diag$(ST, 1)$ in the group GL$(n+1, C(X))$, and thus matrix multiplication is compatible with the equivalence relation that defines GL$(C(X))$. The desired result then follows from Proposition 1.3.10. □

We can think of elements of GL$(C(X))$ as invertible countably infinite matrices with entries in $C(X)$, with all but finitely many of the entries on the main diagonal equal to 1 and all but finitely many off-diagonal entries equal to 0. The identity element I of the group is the infinite matrix with 1s everywhere on the main diagonal and 0s off of it. In particular, while GL$(n, C(X))$ is a subset of M$(n, C(X))$ for all n, GL$(C(X))$ is *not* a subset of M$(C(X))$.

Definition 2.3.3 *For each compact Hausdorff space X, define* K$^{-1}(X)$ *to be the quotient group* GL$(C(X))/$GL$(C(X))_{\mathbf{0}}$.

If S is an element of GL$(C(X))$, we denote its image in K$^{-1}(X)$ by $[\mathsf{S}]$. We use multiplicative notation when working with K^{-1}, so in these cases we often denote the identity element by 1.

The reader may wonder why this group is denoted $K^{-1}(X)$ instead of $K^1(X)$. The reason for this is that K-theory is an example of a *cohomology theory*, and the notational conventions for such a theory dictate the -1 superscript; we shall discuss these issues in the final section of the chapter. However, the antepenultimate section of the chapter will show that we could have written $K^1(X)$ after all.

Example 2.3.4 *Take* S *in* $\mathrm{GL}(C(pt)) \cong \mathrm{GL}(\mathbb{C})$ *and choose* n *large enough so that* S *is an element of* $\mathrm{GL}(n, \mathbb{C})$. *Proposition 1.3.11 states that* S *is in* $\mathrm{GL}(n, \mathbb{C})_{\mathbf{0}}$, *whence* $[\mathsf{S}] = 1$ *in* $K^{-1}(pt)$. *The choice of* S *was arbitrary, and therefore* $K^{-1}(pt) \cong 1$.

Example 2.3.5 *Let* X *be the disjoint union of compact Hausdorff spaces* $X_1, X_2, \ldots, X_k$. *There is an obvious isomorphism*

$$\mathrm{GL}(C(X))/\mathrm{GL}(C(X))_0 \cong \bigoplus_{i=1}^{k} \mathrm{GL}(C(X_i))/\mathrm{GL}(C(X_i))_0,$$

and therefore

$$K^{-1}(X) \cong K^{-1}(X_1) \oplus K^{-1}(X_2) \oplus \cdots \oplus K^{-1}(X_k).$$

Proposition 2.3.6 *Let* X *be compact Hausdorff, let* n *be a natural number, and suppose that* $\{\mathsf{S}_t\}$ *is a homotopy in* $\mathrm{GL}(n, C(X))$. *Then* $[\mathsf{S}_0] = [\mathsf{S}_1]$ *in* $K^{-1}(X)$.

Proof The homotopy $\{\mathsf{S}_t \mathsf{S}_1^{-1}\}$ is a homotopy of invertibles from $\mathsf{S}_0 \mathsf{S}_1^{-1}$ to the identity, whence $\mathsf{S}_0 \mathsf{S}_1^{-1}$ is an element of $\mathrm{GL}(C(X))_0$. Therefore $1 = [\mathsf{S}_0 \mathsf{S}_1^{-1}] = [\mathsf{S}_0][\mathsf{S}_1]^{-1}$, and thus $[\mathsf{S}_0] = [\mathsf{S}_1]$.   □

Corollary 2.3.7 *Let* X *be a compact Hausdorff space and let* n *be a natural number. Then* $[ST] = [\mathrm{diag}(S, T)]$ *in* $K^{-1}(X)$ *for all* S *and* T *in* $\mathrm{GL}(n, C(X))$.

Proof Follows immediately from Propositions 1.3.3 and 2.3.6.   □

Proposition 2.3.8 *The assignment* $X \mapsto K^{-1}(X)$ *is a contravariant functor from the category of compact Hausdorff spaces to the category of abelian groups.*

Proof Suppose $\phi : X \longrightarrow Y$ is a continuous function between compact Hausdorff spaces. Then as we noted in Chapter 1, the map ϕ induces an algebra homomorphism $\phi^* : C(Y) \longrightarrow C(X)$. By applying this homomorphism entrywise, we get a group homomorphism $\phi^* : \mathrm{GL}(C(Y)) \longrightarrow \mathrm{GL}(C(X))$. Take an element S in $\mathrm{GL}(C(Y))$ and choose n large enough so that $\mathsf{S} \in \mathrm{GL}(n, C(Y))$. Suppose that S is in $\mathrm{GL}(n, C(Y))_0$ and let $\{\mathsf{S}_t\}$ be a homotopy in $\mathrm{GL}(n, C(Y))_0$ from $I_n = \mathsf{S}_0$ to $\mathsf{S} = \mathsf{S}_1$. Then $\{\phi^*\mathsf{S}_t\}$ is a homotopy in $\mathrm{GL}(n, C(X))_0$ from I_n to $\phi^*\mathsf{S}$, and thus $\phi^*\mathsf{S}$ is an element of $\mathrm{GL}(C(X))_0$. Therefore ϕ^* descends to a group homomorphism of the quotient groups; i.e., a group homomorphism ϕ^* from K$^{-1}(Y)$ to K$^{-1}(X)$. If id is the identity map on X, then clearly id^* is the identity on K$^{-1}(X)$, and if $\psi : Y \longrightarrow Z$ is a continuous function between compact spaces, we have $(\psi\phi)^* = \phi^*\psi^*$, as desired. $\qquad\square$

Proposition 2.3.9 *Let X and Y be compact Hausdorff spaces.*

   (i) *Suppose ϕ_0 and ϕ_1 are homotopic functions from X to Y. Then ϕ_0^* and ϕ_1^* from K$^{-1}(Y)$ to K$^{-1}(X)$ are equal homomorphisms.*

   (ii) *If X and Y are homotopy equivalent, then* K$^{-1}(X) \cong$ K$^{-1}(Y)$.

Proof Let $\{\phi_t\}$ be a homotopy from ϕ_0 to ϕ_1. Choose S in $\mathrm{GL}(C(X))$ and a natural number n large enough so that S lives in $\mathrm{GL}(n, C(X))$. Then $\{\phi_t^*(\mathsf{S})\}$ is a homotopy in $\mathrm{GL}(n, C(X))$ from $\phi_0^*(\mathsf{S})$ to $\phi_1^*(\mathsf{S})$, and by Lemma 2.3.6, we have $\phi_0^*[\mathsf{S}] = \phi_1^*[\mathsf{S}]$. This proves part (i); part (ii) follows from Proposition 2.3.8. $\qquad\square$

Corollary 2.3.10 *Let X be a contractible compact Hausdorff space. Then* K$^{-1}(X) \cong 1$.

Proof Choose a point x_0 in X so that there is a homotopy from the identity map on X to the constant map on X whose range is $\{x_0\}$. Then K$^{-1}(X) \cong$ K$^{-1}(\{x_0\}) \cong 1$ by Example 2.3.4 and Proposition 2.3.9. $\qquad\square$

Definition 2.3.11 *Let (X, A) be a compact pair. The relative group* K$^{-1}(X, A)$ *is the kernel of the homomorphism $\mu_2^* :$* K$^{-1}(\mathcal{D}(X, A)) \longrightarrow$ K$^{-1}(X)$.

Given an element $(\mathsf{S}_1, \mathsf{S}_2)$ in $\mathrm{GL}(\mathcal{D}(X, A))$, we denote its image in K$^{-1}(\mathcal{D}(X, A))$ as $[\mathsf{S}_1, \mathsf{S}_2]$.

Proposition 2.3.12 *Let (X, A) be a compact pair. Then every element of* $\mathrm{K}^{-1}(X, A)$ *can be written in the form* $[S, I]$ *for some* S *in* $\mathrm{GL}(C(X))$ *that restricts to* I *on* $\mathrm{GL}(C(A))$.

Proof Suppose that $[S_1, S_2]$ is in $\mathrm{K}^{-1}(X, A)$. Then $\mu_2^*[S_1, S_2] = [S_2] = 1$, and so by choosing a sufficiently large natural number n we can take S_1 to be in $\mathrm{GL}(n, C(X))$ and S_2 to be in $\mathrm{GL}(n, C(X))_{\mathbf{0}}$. Choose a homotopy $\{R_t\}$ in $\mathrm{GL}(n, C(X))_{\mathbf{0}}$ from $R_0 = I_n$ to $R_1 = S_2^{-1}$ and set $S = S_2^{-1}S_1$. Then $\{(R_tS_1, R_tS_2)\}$ is a homotopy in $\mathrm{GL}(n, C(\mathcal{D}(X, A)))$ from (S_1, S_2) to (S, I_n), whence $[S, I] = [S_1, S_2]$ in $\mathrm{K}^{-1}(X, A)$. $\qquad\square$

Proposition 2.3.13 *Relative* K^{-1} *defines a contravariant functor from the category of compact pairs to the category of abelian groups.*

Proof The proof of this result is the same as that of Proposition 2.2.7, but with K^0 replaced with K^{-1} everywhere. $\qquad\square$

Theorem 2.3.14 *Let (X, A) be a compact pair and let $j : A \longrightarrow X$ be the inclusion map. Then the sequence*

$$\mathrm{K}^{-1}(X, A) \xrightarrow{\;\mu_1^*\;} \mathrm{K}^{-1}(X) \xrightarrow{\;j^*\;} \mathrm{K}^{-1}(A)$$

is exact.

Proof Suppose $[S, I]$ is in $\mathrm{K}^{-1}(X, A)$. Then

$$j^*\mu_1^*[S, I] = j^*[S] = [S|A] = 1.$$

Now suppose that $[T]$ is in $\mathrm{K}^{-1}(X)$ and that $j^*[T] = [T|A] = 1$ in $\mathrm{K}^{-1}(A)$. For n sufficiently large, we may view T as an element of $\mathrm{GL}(n, C(X))$ and $T|A$ as an element of $\mathrm{GL}(n, C(A))_{\mathbf{0}}$. Proposition 1.3.16 implies that there exists $\widetilde{T}$ in $\mathrm{GL}(n, C(X))_{\mathbf{0}}$ such that $\widetilde{T}$ restricts to $T|A$ on A. Then $[T\widetilde{T}^{-1}, I]$ is an element in $\mathrm{K}^{-1}(X, A)$ and

$$\mu_1^*[T\widetilde{T}^{-1}, I] = [T\widetilde{T}^{-1}] = [T][\widetilde{T}^{-1}] = [T].$$

$$\square$$

Theorem 2.3.15 *Let (X, A) be a compact pair, let $j : A \longrightarrow X$ be the inclusion map, and suppose that $\psi : X \longrightarrow A$ is a continuous function such that ψj is the identity map on A. Then there exists a split exact* *sequence*

$$1 \longrightarrow K^{-1}(X, A) \xrightarrow{\ \mu_1^* \ } K^{-1}(X) \underset{\psi^*}{\overset{j^*}{\rightleftarrows}} K^{-1}(A) \longrightarrow 1,$$

whence $K^{-1}(X)$ *is isomorphic to* $K^{-1}(A) \oplus K^{-1}(X, A)$.

Proof As in the proof of Theorem 2.2.10, the only nonobvious part of the theorem is exactness at $K^{-1}(X, A)$. Take $[S, I]$ in $K^{-1}(X, A)$, and suppose that $\mu_1^*[S, I] = [S] = 1$ in $K^{-1}(X)$. Then S is in $GL(n, C(X))_0$ for some natural number n; choose a homotopy $\{S_t\}$ in $GL(n, C(X))_0$ from $S_0 = I_n$ to $S_1 = S$. Then

$$(\psi^* j^* S_t)|A = j^* \psi^* j^* S_t = j^* S_t = S_t|A$$

for all $0 \le t \le 1$, and thus $(S_t, \psi^* j^* S_t)$ is in $GL(n, C(\mathcal{D}(X, A)))$. Note that $\psi^* j^* S_0 = \psi^* j^* I_n = I_n$ and $\psi^* j^* S_1 = \psi^* I_n = I_n$. Therefore $\{(S_t, \psi^* j^* S_t)\}$ is a homotopy in $GL(n, C(\mathcal{D}(X, A)))$ from (I_n, I_n) to (S, I_n), and thus $[S, I]$ is the identity element in $K^{-1}(X, A)$. ☐

Corollary 2.3.16 *Let X be a compact Hausdorff space and take x_0 in X. Then* $K^{-1}(X) \cong K^{-1}(X, x_0)$.

Proof The sequence

$$1 \longrightarrow K^{-1}(X, x_0) \xrightarrow{\ \mu_1^* \ } K^{-1}(X) \underset{\psi^*}{\overset{j^*}{\rightleftarrows}} K^{-1}(\{x_0\}) \longrightarrow 1$$

is split exact, and $K^1(\{x_0\})$ is trivial by Example 2.3.4. ☐

Theorem 2.3.17 *Let (X, A) be a compact pair, and let $q : (X, A) \longrightarrow ((X \backslash A)^+, \infty)$ be the quotient morphism. Then*

$$q^* : K^{-1}((X \backslash A)^+, \infty) \longrightarrow K^{-1}(X, A)$$

is an isomorphism. Furthermore, suppose that $\phi : (X, A) \longrightarrow (Y, B)$ is a morphism of compact pairs, and define $\widetilde{\phi} : (X \backslash A)^+ \longrightarrow (Y \backslash B)^+$ by the formula

$$\widetilde{\phi}(z) = \begin{cases} z & \text{if } z \in X \backslash A \\ \infty & \text{if } z = \infty. \end{cases}$$

Then the diagram

$$K^{-1}((Y\backslash B)^+,\infty) \xrightarrow{\;\widetilde{\phi}^*\;} K^{-1}((X\backslash A)^+,\infty)$$

$$\downarrow q^* \qquad\qquad\qquad\qquad \downarrow q*$$

$$K^{-1}(Y,B) \xrightarrow{\;\phi^*\;} K^{-1}(X,A)$$

commutes.

Proof Take $[S,I]$ in $K^{-1}(X,A)$ and choose a natural number n large enough so that S is in $GL(n,C(X))$. By Lemma 2.2.13, there exists $\widehat{S}$ in $GL(n,C((X\backslash A)^+)$ such that $q^*\widehat{S} = S$. Therefore

$$q^*[\widehat{S},I] = [q^*\widehat{S},I] = [S,I],$$

and thus q^* is a surjection.

To show that q^* is an injection, suppose that $[R,I]$ in $K^{-1}((X\backslash A)^+,\infty)$ and that $q^*[R,I] = [q^*R,I] = 1$ in $K^{-1}(X,A)$. For some natural number n, there exists a homotopy $\{(S_t,\widetilde{S}_t)\}$ in $GL(n,C(\mathcal{D}(X,A)))_0$ such that $(S_0,\widetilde{S}_0) = (I_n,I_n)$ and $(S_1,\widetilde{S}_1) = (q^*R,I_n)$. Moreover, $\{(S_t\widetilde{S}_t^{-1},I_n)\}$ is also a homotopy in $GL(n,C(\mathcal{D}(X,A)))_0$ from (I_n,I_n) to (q^*R,I_n). By Lemma 2.2.13, there exists for each t a unique $\widetilde{T}_t$ in $M(n,C((X\backslash A)^+))$ with the property that $q^*\widetilde{T}_t = S_t\widetilde{S}_t^{-1}$. The proof of Lemma 2.2.13 shows that each T_t is invertible and that $\{T_t\}$ varies continuously as a function of t. Therefore $\{(\widetilde{T}_t,I_n)\}$ is a homotopy in $GL(n,C(\mathcal{D}((X\backslash A)^+,\infty)))$ from $(\widetilde{T}_0,I_n) = (I_n,I_n)$ to $(\widetilde{T}_1,I_n) = (R,I_n)$, and thus $[R,I] = 1$ in $K^{-1}((X\backslash A)^+,\infty)$. Finally, a direct calculation shows that the diagram in the statement of the theorem commutes. $\qquad\square$

Corollary 2.3.18 (Excision for K^{-1}) *Let (X,A) be a compact pair and let U be an open subspace of A. Then there exists a natural isomorphism*

$$K^{-1}(X\backslash U, A\backslash U) \cong K^{-1}(X,A).$$

Proof Let

$$q_1 : (X,A) \longrightarrow ((X\backslash A)^+,\infty)$$

and

$$q_2 : (X\backslash U, A\backslash U) \longrightarrow (((X\backslash U)\backslash(A\backslash U))^+,\infty)$$

be the quotient maps described in Definition 2.2.12. By Theorem 2.3.17, the maps q_1^* and q_2^* are isomorphisms. Then because $(X\backslash U)\backslash(A\backslash U) =$

$X \backslash A$, the composition $q_1^*(q_2^*)^{-1}$ is well defined and is a natural isomorphism.   □

2.4 Connecting K^0 and K^{-1}

In this section we show how the short exact sequences in Theorems 2.2.9 and 2.3.14 can be combined via a *connecting homomorphism* to form a longer exact sequence; as we will see, constructing the connecting homomorphism and verifying its properties is a process that is quite involved. Later in the chapter we will see that our two short exact sequences can be connected in another way that is far less obvious.

Theorem 2.4.1 *Let* (X, A) *be a compact pair. There is a natural homomorphism* ∂ *from* K⁻¹(A) *to* K⁰(X, A) *that makes the following sequence exact:*

$$K^{\text{-}1}(X, A) \xrightarrow{\;\mu_1^*\;} K^{\text{-}1}(X) \xrightarrow{\;j^*\;} K^{\text{-}1}(A) \xrightarrow{\;\partial\;}$$

$$K^0(X, A) \xrightarrow{\;\mu_1^*\;} K^0(X) \xrightarrow{\;j^*\;} K^0(A).$$

Proof The proof of this theorem is long and involved, so we break it into pieces.

Definition of ∂:

Take S in $GL(n, C(A))$ and choose T in $GL(2n, C(X))$ with the property that $\mathsf{T}|A = \text{diag}(\mathsf{S}, \mathsf{S}^{-1})$; Corollary 1.3.17 guarantees the existence of such a T. Define

$$\partial[\mathsf{S}] = \left[\mathsf{T} \, \text{diag}(I_n, 0_n) \mathsf{T}^{-1}, \text{diag}(I_n, 0_n) \right] - [I_n, I_n].$$

∂ is well defined:

Choose $\widetilde{\mathsf{T}}$ in $GL(2n, C(X))$ with the property that $\widetilde{\mathsf{T}}$ restricts to $\text{diag}(\mathsf{S}, \mathsf{S}^{-1})$ on A. Then $\widetilde{\mathsf{T}} \mathsf{T}^{-1}|A = I_{2n}$, and thus $(\widetilde{\mathsf{T}} \mathsf{T}^{-1}, I_{2n})$ is in $GL(2n, C(\mathcal{D}(X, A)))$. We have

$$\left(\widetilde{\mathsf{T}} \mathsf{T}^{-1}, I_{2n}\right) \left(\mathsf{T} \, \text{diag}(I_n, 0_n) \mathsf{T}^{-1}, \text{diag}(I_n, 0_n)\right) \left(\widetilde{\mathsf{T}} \mathsf{T}^{-1}, I_{2n}\right)^{-1}$$
$$= \left(\widetilde{\mathsf{T}} \, \text{diag}(I_n, 0_n) \widetilde{\mathsf{T}}^{-1}, \text{diag}(I_n, 0_n)\right),$$

which implies that $\partial[\mathsf{S}]$ does not depend on our choice of T.

Second, we know from Lemma 1.7.1 that there exists an element U

in $\mathrm{GL}(2(n+1), \mathbb{C})$ such that $\mathrm{diag}(\mathsf{S}, 1, \mathsf{S}^{-1}, 1) = \mathsf{U}\,\mathrm{diag}(\mathsf{S}, \mathsf{S}^{-1}, I_2)\mathsf{U}^{-1}$. Then $\mathsf{U}\,\mathrm{diag}(\mathsf{T}, I_2)\mathsf{U}^{-1}$ is an element of $\mathrm{GL}(2(n+1), C(X))$ with the property that

$$(\mathsf{U}\,\mathrm{diag}(\mathsf{T}, I_2)\mathsf{U}^{-1}) \mid A = \mathrm{diag}(\mathsf{S}, 1, \mathsf{S}^{-1}, 1)$$

and

$$\mathrm{diag}(I_{n+1}, 0_{n+1}) = \\ ((\mathsf{U}\,\mathrm{diag}(\mathsf{T}, I_2)\mathsf{U}^{-1})\,\mathrm{diag}(I_{n+1}, 0_{n+1})(\mathsf{U}\,\mathrm{diag}(\mathsf{T}, I_2)\mathsf{U}^{-1})^{-1}) \mid A.$$

Employing Lemma 2.2.4, we have

$$\begin{aligned}
&\left[(\mathsf{U}\,\mathrm{diag}(\mathsf{T}, I_2)\mathsf{U}^{-1})\,\mathrm{diag}(I_{n+1}, 0_{n+1})(\mathsf{U}\,\mathrm{diag}(\mathsf{T}, I_2)\mathsf{U}^{-1})^{-1},\right.\\
&\qquad\qquad\qquad\qquad\qquad \left.\mathrm{diag}(I_{n+1}, 0_{n+1})\right] - [I_{n+1}, I_{n+1}]\\
={}&\left[\mathsf{U}\,\mathrm{diag}(\mathsf{T}, I_2)\,\mathrm{diag}(I_n, 0_n, 1, 0)\,\mathrm{diag}(\mathsf{T}, I_2)^{-1}\mathsf{U}^{-1},\right.\\
&\qquad\qquad\qquad\qquad\qquad \left.\mathrm{diag}(I_{n+1}, 0_{n+1})\right] - [I_{n+1}, I_{n+1}]\\
={}&\left[\mathrm{diag}(\mathsf{T}, I_2)\,\mathrm{diag}(I_n, 0_n, 1, 0)\,\mathrm{diag}(\mathsf{T}, I_2)^{-1},\right.\\
&\qquad\qquad\qquad\qquad \left.\mathsf{U}^{-1}\,\mathrm{diag}(I_{n+1}, 0_{n+1})\mathsf{U}\right] - [I_{n+1}, I_{n+1}]\\
={}&\left[\mathrm{diag}(\mathsf{T}, I_2)\,\mathrm{diag}(I_n, 0_n, 1, 0)\,\mathrm{diag}(\mathsf{T}, I_2)^{-1},\right.\\
&\qquad\qquad\qquad\qquad \left.\mathrm{diag}(I_n, 0_n, 1, 0)\right] - [I_{n+1}, I_{n+1}]\\
={}&[\mathsf{T}\,\mathrm{diag}(I_n, 0_n)\mathsf{T}^{-1}, \mathrm{diag}(I_n, 0_n)] + [\mathrm{diag}(1, 0), \mathrm{diag}(1, 0)]\\
&\qquad\qquad\qquad\qquad\qquad - [\mathrm{diag}(I_n, 1), \mathrm{diag}(I_n, 1)]\\
={}&[\mathsf{T}\,\mathrm{diag}(I_n, 0_n)\mathsf{T}^{-1}, \mathrm{diag}(I_n, 0_n)] + [\mathrm{diag}(0, 1), \mathrm{diag}(0, 1)]\\
&\qquad\qquad\qquad\qquad\qquad - [\mathrm{diag}(I_n, 1), \mathrm{diag}(I_n, 1)]\\
={}&[\mathsf{T}\,\mathrm{diag}(I_n, 0_n)\mathsf{T}^{-1}, \mathrm{diag}(I_n, 0_n)] - [I_n, I_n],
\end{aligned}$$

and thus $\partial[\mathrm{diag}(\mathsf{S}, 1)] = \partial[\mathsf{S}]$. Therefore $\partial[\mathsf{S}]$ does not depend on the size of the matrix we use to represent $[\mathsf{S}]$.

Third, suppose that $[\mathsf{S}_0] = [\mathsf{S}_1]$. We suppose that n is sufficiently large so that S_0 and S_1 are in the same connected component of $\mathrm{GL}(n, C(A))$. Choose a homotopy $\{\mathsf{S}_t\}$ in $\mathrm{GL}(n, C(A))$ from S_0 to S_1 and define $\widehat{S}$ in $\mathrm{GL}(n, C(A \times [0,1]))$ by the formula $\widehat{S}(a, t) = \mathsf{S}_t(a)$. Use Corollary 1.3.17 to choose $\widehat{T}$ in $\mathrm{GL}(2n, C(X \times [0,1]))$ with the property that $\widehat{T}|(A \times [0,1]) = \mathrm{diag}(\widehat{S}, \widehat{S}^{-1})$, and for each x in X and $0 \leq t \leq 1$, set $\mathsf{T}_t(x) = \widehat{T}(x, t)$. Then $\{(\mathsf{T}_t\,\mathrm{diag}(I_n, 0_n)\mathsf{T}_t^{-1}, \mathrm{diag}(I_n, 0_n))\}$ is a homotopy in $\mathrm{GL}(2n, C(\mathcal{D}(X, A)))$, and therefore

$$\begin{aligned}
\partial[\mathsf{S}_0] &= [\mathsf{T}_0\,\mathrm{diag}(I_n, 0_n)\mathsf{T}_0^{-1}, \mathrm{diag}(I_n, 0_n)] - [I_n, I_n]\\
&= [\mathsf{T}_1\,\mathrm{diag}(I_n, 0_n)\mathsf{T}_1^{-1}, \mathrm{diag}(I_n, 0_n)] - [I_n, I_n] = \partial[\mathsf{S}_1].
\end{aligned}$$

∂ is a group homomorphism:

Take $[S_1]$ and $[S_2]$ in $K^{-1}(A)$ and choose n large enough so that we may view S_1 and S_2 as elements of $GL(n, C(A))$. For $i = 1, 2$, choose T_i in $GL(2n, C(X))$ so that $T_i|A = \text{diag}(S_i, S_i^{-1})$. By Lemma 1.7.1, there exists a matrix U in $GL(4n, \mathbb{C})$ with the properties that

$$U \, \text{diag}(S_1, S_1^{-1}, S_2, S_2^{-1}) U^{-1} = \text{diag}(S_1, S_2, S_1^{-1}, S_2^{-1})$$
$$U \, \text{diag}(I_n, 0_n, I_n, 0_n) U^{-1} = \text{diag}(I_{2n}, 0_{2n}).$$

Then

$$\left[(U \, \text{diag}(T_1, T_2) U^{-1}) \, \text{diag}(I_{2n}, 0_{2n})(U \, \text{diag}(T_1, T_2) U^{-1})^{-1}, \right.$$
$$\left. \text{diag}(I_{2n}, 0_{2n}) \right]$$
$$= \left[\text{diag}(T_1, T_2) \, \text{diag}(I_n, 0_n, I_n, 0_n) \, \text{diag}(T_1, T_2)^{-1}, \right.$$
$$\left. U^{-1} \, \text{diag}(I_{2n}, 0_{2n}) U \right]$$
$$= \left[\text{diag}(T_1, T_2) \, \text{diag}(I_n, 0_n, I_n, 0_n) \, \text{diag}(T_1, T_2)^{-1}, \right.$$
$$\left. \text{diag}(I_n, 0_n, I_n, 0_n) \right]$$
$$= [T_1 \, \text{diag}(I_n, 0_n) T_1^{-1}, \text{diag}(I_n, 0_n)] +$$
$$[T_2 \, \text{diag}(I_n, 0_n) T_2^{-1}, \text{diag}(I_n, 0_n)],$$

and Corollary 2.3.7 gives us

$$\partial[S_1 S_2] = \partial[\text{diag}(S_1, S_2)] = \partial[S_1] + \partial[S_2].$$

Exactness at $K^{-1}(A)$:

Take R in $GL(n, C(X))$. Then

$$\partial j^*[R] = [R|A]$$
$$= [\text{diag}(R, R^{-1}) \, \text{diag}(I_n, 0_n) \, \text{diag}(R, R^{-1})^{-1}, \text{diag}(I_n, 0_n)] - [I_n, I_n]$$
$$= [\text{diag}(I_n, 0_n), \text{diag}(I_n, 0_n)] - [I_n, I_n] = 0.$$

Now suppose that

$$\partial[S] = [T \, \text{diag}(I_n, 0_n) T^{-1}, \text{diag}(I_n, 0_n)] - [I_n, I_n] = 0$$

in $K^0(X, A)$ for some S in $GL(n, C(A))$. From Proposition 2.1.5 and the definition of $\mathcal{D}(X, A)$, there exist a natural number m and elements V_1 and V_2 in $GL(2(n + m), C(X))$ such that:

- $V_1 \, \text{diag}\left(T \, \text{diag}(I_n, 0_n) T^{-1}, I_m, 0_m\right) V_1^{-1} = \text{diag}(I_{n+m}, 0_{n+m});$
- $V_2 \, \text{diag}(I_n, 0_n, I_m, 0_m) V_2^{-1} = \text{diag}(I_{n+m}, 0_{n+m});$
- $V_1(a) = V_2(a)$ for all a in A.

Lemma 1.7.1 gives us a matrix U in $\mathrm{GL}(2(n+m), \mathbb{C})$ with the property that

$$U \operatorname{diag}(I_{n+m}, 0_{n+m}) U^{-1} = \operatorname{diag}(I_n, 0_n, I_m, 0_m),$$

and thus

$$
\begin{aligned}
\operatorname{diag}&(I_{n+m}, 0_{n+m}) \\
&= V_1 \operatorname{diag}(T, I_{2m}) \operatorname{diag}(I_n, 0_n, I_m, 0_m) \operatorname{diag}(T, I_{2m})^{-1} V_1^{-1} \\
&= V_1 \operatorname{diag}(T, I_{2m}) U \operatorname{diag}(I_{n+m}, 0_{n+m}) U^{-1} \operatorname{diag}(T, I_{2m})^{-1} V_1^{-1}.
\end{aligned}
$$

We also know that

$$
\begin{aligned}
\operatorname{diag}(I_{n+m}, 0_{n+m}) &= V_2 \operatorname{diag}(I_n, 0_n, I_m, 0_m) V_2^{-1} \\
&= V_2 U \operatorname{diag}(I_{n+m}, 0_{n+m}) U^{-1} V_2^{-1}.
\end{aligned}
$$

Therefore

$$
\begin{aligned}
V_1 \operatorname{diag}(T, I_{2m}) U \operatorname{diag}(I_{n+m}, 0_{n+m}) U^{-1} \operatorname{diag}(T, I_{2m})^{-1} V_1^{-1} = \\
V_2 U \operatorname{diag}(I_{n+m}, 0_{n+m}) U^{-1} V_2^{-1},
\end{aligned}
$$

and hence

$$
\begin{aligned}
\left(U^{-1} V_2^{-1} V_1 \operatorname{diag}(T, I_{2m}) U\right) \operatorname{diag}(I_{n+m}, 0_{n+m}) = \\
\operatorname{diag}(I_{n+m}, 0_{n+m}) \left(U^{-1} V_2^{-1} V_1 \operatorname{diag}(T, I_{2m}) U\right).
\end{aligned}
$$

By Proposition 1.3.18, there exist V_{11} and V_{22} in $\mathrm{GL}(n+m, C(X))$ such that

$$U^{-1} V_2^{-1} V_1 \operatorname{diag}(T, I_{2m}) U = \operatorname{diag}(V_{11}, V_{22}).$$

For all a in A, we have

$$
\begin{aligned}
U^{-1} V_2^{-1}(a) V_1(a) \operatorname{diag}(T(a), I_{2m}) U &= U^{-1} \operatorname{diag}(S(a), S^{-1}(a), I_{2m}) U \\
&= \operatorname{diag}(S(a), I_m, S^{-1}(a), I_m),
\end{aligned}
$$

and so $V_{11}|A = \operatorname{diag}(S, I_m)$. Hence

$$j^*[V_{11}] = [\operatorname{diag}(S, I_m)] = [S][I_m] = [S].$$

Exactness at $K^0(X, A)$:

Take S in $GL(n, C(A))$ and choose T in $GL(2n, C(X))$ with $T|_A = \text{diag}(S, S^{-1})$. Then

$$\begin{aligned}
\mu_1^* \partial[S] &= \mu_1^* \left([T \, \text{diag}(I_n, 0_n) T^{-1}, \text{diag}(I_n, 0_n)] - [I_n, I_n] \right) \\
&= [T \, \text{diag}(I_n, 0_n) T^{-1}] - [I_n] \\
&= [\text{diag}(I_n, 0_n)] - [I_n] \\
&= 0.
\end{aligned}$$

Now choose $[E, \text{diag}(I_n, 0_n)] - [I_n, I_n]$ in $K^0(X, A)$ and suppose that

$$\mu_1^* \left([E, \text{diag}(I_n, 0_n)] - [I_n, I_n] \right) = [E] - [I_n] = 0$$

in $K^0(X)$. Thanks to Propositions 1.4.10 and 2.1.5, we know that there exists a natural number l with the property that

$$\text{diag}(E, I_l, 0_l) \sim_h \text{diag}(I_{n+l}, 0_{n+l}).$$

Choose a homotopy $\{F(t)\}$ of idempotents in $M(2(n+l), C(X))$ from $\text{diag}(I_{n+l}, 0_{n+l})$ to $\text{diag}(E, I_l, 0_l)$; for notational convenience in the rest of this proof, we will write our homotopies in this form instead of the usual $\{F_t\}$. Set

$$M = \sup \left\{ \left\| 2F(t) - I_{2(n+l)} \right\|_\infty : t \in [0, 1] \right\},$$

and choose points $0 = t_0 < t_1 < t_2 < \cdots < t_K = 1$ so that

$$\| F(t_{k-1}) - F(t_k) \|_\infty < \frac{1}{M}$$

for all $1 \le k \le K$. For each k, define

$$\widetilde{V}_k(t) = I_{2(n+l)} - F(t) - F(t_{k-1}) + 2F(t)F(t_{k-1}).$$

By Lemma 1.4.8, the matrix $\widetilde{V}_k(t)$ is invertible when $t_{k-1} \le t \le t_k$, and

$$\widetilde{V}_k(t_k) F(t_{k-1}) \widetilde{V}_k^{-1}(t_k) = F(t_k).$$

Define

$$\widetilde{V}(t) = \begin{cases}
\widetilde{V}_1(t) & 0 \le t \le t_1 \\
\widetilde{V}_2(t)\widetilde{V}_1(t_1) & t_1 \le t \le t_2 \\
\widetilde{V}_3(t)\widetilde{V}_2(t_2)\widetilde{V}_1(t_1) & t_2 \le t \le t_3 \\
\cdots & \cdots \\
\widetilde{V}_K(t)\widetilde{V}_{K-1}(t_{K-1}) \cdots \widetilde{V}_1(t_1) & t_{K-1} \le t \le 1.
\end{cases}$$

Note that

$$\widetilde{V}_k(t_{k-1}) = I_{2(n+l)} - F(t_{k-1}) - F(t_{k-1}) + 2F(t_{k-1})F(t_{k-1}) = I_{2(n+l)}$$

for each $1 \le k \le K$, which implies that $\widetilde{V}$ is well defined and continuous. Set $V = \widetilde{V}(1)$. Then

$$V \operatorname{diag}(I_{n+l}, 0_{n+l})V^{-1} = VF(t_0)V^{-1} = F(t_1) = \operatorname{diag}(E, I_l, 0_l).$$

Choose U in $GL(2(n+l), \mathbb{C})$ so that

$$U \operatorname{diag}(I_n, 0_n, I_l, 0_l)U^{-1} = \operatorname{diag}(I_{n+l}, 0_{n+l}).$$

Then

$$[E, \operatorname{diag}(I_n, 0_n)] - [I_n, I_n]$$
$$= [\operatorname{diag}(E, I_l, 0_l), \operatorname{diag}(I_n, 0_n, I_l, 0_l] - [I_{n+l}, I_{n+l}]$$
$$= [V \operatorname{diag}(I_{n+l}, 0_{n+l})V^{-1}, U^{-1} \operatorname{diag}(I_{n+l}, 0_{n+l})U] - [I_{n+l}, I_{n+l}]$$
$$= [(UV) \operatorname{diag}(I_{n+l}, 0_{n+l})(UV)^{-1}, \operatorname{diag}(I_{n+l}, 0_{n+l})] - [I_{n+l}, I_{n+l}].$$

However, we are not done, because is it is not necessarily true that $(UV)|A = \operatorname{diag}(S, S^{-1})$ for some S in $GL(n+l, C(A))$.

We need to modify UV in some fashion. Note that we can not only take U to be an element of $GL(2(n+l), \mathbb{C})$, but also an element of $GL(2(n+l), C(X))$ with each matrix entry constant. Because $K^{-1}(pt)$ is trivial, there is a homotopy in $GL(C(pt))$ from U to I, and this homotopy also lives in $GL(C(X))$, whence $[U] = 1$ in $K^{-1}(X)$. Our construction of V shows that $[V] = 1$ in $K^{-1}(X)$, and thus $[UV] = [U][V] = [I][I] = 1$ in $K^{-1}(X)$. Therefore $[(UV)|A] = 1$ in $K^{-1}(A)$ as well. We have

$$((UV)|A) \operatorname{diag}(I_{n+l}, 0_{n+l})((UV)|A)^{-1} = U \operatorname{diag}(I_n, 0_n, I_l, 0_l)U^{-1}$$
$$= \operatorname{diag}(I_{n+l}, 0_{n+l}),$$

so Proposition 1.3.18 implies that $(UV)|A = \operatorname{diag}(W_{11}, W_{22})$ for some W_{11} and W_{22} in $GL(n+l, C(A))$. Putting these two facts together, we obtain

$$1 = 1^{-1} = [\operatorname{diag}(W_{11}, W_{22})]^{-1} = [W_{11}W_{22}]^{-1} = [W_{22}^{-1}W_{11}^{-1}]$$

in $K^{-1}(A)$. Thus for a sufficiently large natural number N, the matrix $\operatorname{diag}(W_{22}^{-1}W_{11}^{-1}, I_N)$ is in $GL(n+l+N, C(A))_0$. By Proposition 1.3.16 we know that $\operatorname{diag}(W_{22}^{-1}W_{11}^{-1}, I_N)$ is the restriction to A of an element W in $GL(n+l+N, C(X))$. Use Lemma 1.7.1 to choose J in $GL(2(n+l+N), \mathbb{C})$

such that

$$\mathsf{J}\operatorname{diag}(\mathsf{W}_{11}, \mathsf{W}_{22}, I_N, I_N)\mathsf{J}^{-1} = \operatorname{diag}(\mathsf{W}_{11}, I_N, \mathsf{W}_{11}^{-1}, I_N)$$
$$\mathsf{J}\operatorname{diag}(I_{n+l}, 0_{n+l}, I_N, 0_N)\mathsf{J}^{-1} = \operatorname{diag}(I_{n+l+N}, 0_{n+l+N}),$$

and define

$$\mathsf{L} = \mathsf{J}\operatorname{diag}(\mathsf{UV}, I_N, I_N)\mathsf{J}^{-1}\operatorname{diag}(I_{n+l+N}, \mathsf{W}).$$

Then

$$\mathsf{L}|A = \mathsf{J}\operatorname{diag}(\mathsf{W}_{11}, \mathsf{W}_{22}, I_N, I_N)\mathsf{J}^{-1}\operatorname{diag}(I_{n+l+N}, \mathsf{W}_{22}^{-1}\mathsf{W}_{11}^{-1}, I_N)$$
$$= \operatorname{diag}(\mathsf{W}_{11}, I_N, \mathsf{W}_{22}, I_N)\operatorname{diag}(I_{n+l+N}, \mathsf{W}_{22}^{-1}\mathsf{W}_{11}^{-1}, I_N)$$
$$= \operatorname{diag}(\mathsf{W}_{11}, I_N, \mathsf{W}_{11}^{-1}, I_N),$$

which implies that

$$\left[\mathsf{L}\operatorname{diag}(I_{n+l+N}, 0_{n+l+N})\mathsf{L}^{-1}, \operatorname{diag}(I_{n+l+N}, 0_{n+l+N})\right]$$

is an element of $\mathrm{K}^0(\mathcal{D}(X, A))$. The definition of L yields

$$\mathsf{L}\operatorname{diag}(I_{n+l+N}, 0_{n+l+N})\mathsf{L}^{-1}$$
$$= \mathsf{J}\operatorname{diag}(\mathsf{UV}, I_N, I_N)\mathsf{J}^{-1}\operatorname{diag}(I_{n+l+N}, \mathsf{W})\operatorname{diag}(I_{n+l+N}, 0_{n+l+N})\times$$
$$\qquad\qquad \operatorname{diag}(I_{n+l+N}, \mathsf{W})^{-1}\mathsf{J}\operatorname{diag}(\mathsf{UV}, I_N, I_N)^{-1}\mathsf{J}^{-1}$$
$$= \mathsf{J}\operatorname{diag}(\mathsf{UV}, I_N, I_N)\mathsf{J}^{-1}\operatorname{diag}(I_{n+l+N}, 0_{n+l+N})\mathsf{J}\operatorname{diag}(\mathsf{UV}, I_N, I_N)^{-1}\mathsf{J}^{-1}$$
$$= \mathsf{J}\operatorname{diag}(\mathsf{UV}, I_N, I_N)\operatorname{diag}(I_{n+l}, 0_{n+l}, I_N, 0_N)\operatorname{diag}(\mathsf{UV}, I_N, I_N)^{-1}\mathsf{J}^{-1}$$
$$= \mathsf{J}\big(\operatorname{diag}((\mathsf{UV})\operatorname{diag}(I_{n+l}, 0_{n+l})(\mathsf{UV})^{-1}, I_N, 0_N)\big)\mathsf{J}^{-1},$$

and Lemma 2.2.4 implies that

$$\left[\mathsf{L}\operatorname{diag}(I_{n+l+N}, 0_{n+l+N})\mathsf{L}^{-1}, \operatorname{diag}(I_{n+l+N}, 0_{n+l+N})\right]$$
$$= \big[\operatorname{diag}((\mathsf{UV})\operatorname{diag}(I_{n+l}, 0_{n+l})(\mathsf{UV})^{-1}, I_N, 0_N),$$
$$\qquad\qquad\qquad \mathsf{J}^{-1}\operatorname{diag}(I_{n+l+N}, 0_{n+l+N})\mathsf{J}\big]$$
$$= \big[\operatorname{diag}((\mathsf{UV})\operatorname{diag}(I_{n+l}, 0_{n+l})(\mathsf{UV})^{-1}, I_N, 0_N),$$
$$\qquad\qquad\qquad \operatorname{diag}(I_{n+l}, 0_{n+l}, I_N, 0_N)\big]$$
$$= \big[\mathsf{UV}\operatorname{diag}(I_{n+l}, 0_{n+l})(\mathsf{UV})^{-1}, \operatorname{diag}(I_{n+l}, 0_{n+l})\big] + [I_N, I_N].$$

Therefore

$$\partial[\mathrm{diag}(W_{11}, I_N)]$$
$$= \left[\mathsf{L}\,\mathrm{diag}(I_{n+l+N}, 0_{n+l+N})\mathsf{L}^{-1}, \mathrm{diag}(I_{n+l+N}, 0_{n+l+N})\right]$$
$$\qquad\qquad\qquad\qquad\qquad - [I_{n+l+N}, I_{n+l+N}]$$
$$= \left[(\mathsf{UV})\,\mathrm{diag}(I_{n+l}, 0_{n+l})(\mathsf{UV})^{-1}, \mathrm{diag}(I_{n+l}, 0_{n+l})\right] + [I_N, I_N]$$
$$\qquad\qquad\qquad\qquad\qquad - [I_{n+l}, I_{n+l}] - [I_N, I_N]$$
$$= [\mathsf{E}, \mathrm{diag}(I_n, 0_n)] - [I_n, I_n].$$

∂ *is a natural transformation:*

Let $\phi : (X, A) \longrightarrow (Y, B)$ be a morphism of compact pairs and take S in $\mathrm{GL}(n, C(B))$. Corollary 1.3.17 states that we may choose T in $\mathrm{GL}(2n, C(Y))$ with the property that $\mathsf{T}|B = \mathrm{diag}(\mathsf{S}, \mathsf{S}^{-1})$. Then

$$(\phi^*\mathsf{T})(a) = \mathsf{T}(\phi(a)) = \mathrm{diag}\big(\mathsf{S}(\phi(a)), \mathsf{S}^{-1}(\phi(a))\big)$$
$$= \mathrm{diag}\big((\phi^*\mathsf{S})(a), (\phi^*\mathsf{S}^{-1})(a)\big)$$

for all a in A, whence $\partial[\phi^*\mathsf{S}] = \phi^*(\partial[\mathsf{S}])$.   □

2.5 Reduced K-theory

For every compact Hausdorff space X, the group $\mathrm{K}^0(X)$ contains a summand isomorphic to $\mathbb{Z}$; this is the subgroup of $\mathrm{K}^0(X)$ generated by the trivial bundles over X. Therefore, if we strip off this copy of $\mathbb{Z}$, we are left with the part of $\mathrm{K}^0(X)$ that only detects the nontrivial (and hence presumably more interesting) vector bundles over X. *Reduced K-theory* is the way to make this idea precise.

Definition 2.5.1 *Let x_0 be a point in a compact Hausdorff space X. The compact pair (X, x_0) is called a* pointed space, *and x_0 is called the* base point. *A* morphism *from (X, x_0) to a pointed space (Y, y_0) is a continuous function from X to Y such that $\phi(x_0) = y_0$.*

When the particular choice of basepoint in a pointed space (X, x_0) is not important or is clear from the context, we will write just X.

Definition 2.5.2 *Let (X, x_0) be a pointed space and let j denote the inclusion of x_0 into X. For $p = 0, -1$, the reduced group $\widetilde{\mathrm{K}}^{\mathrm{p}}(X)$ is the kernel of homomorphism $j^* : \mathrm{K}^{\mathrm{p}}(X) \longrightarrow \mathrm{K}^{\mathrm{p}}(x_0)$.*

A pointed space is a special case of a compact pair; in this case, the reduced and relative groups are very closely related.

Proposition 2.5.3 *Let (X, x_0) be a pointed space. Then for $p = 0, -1$, there exists a natural isomorphism $\widetilde{K}^p(X, x_0) \cong K^p(X, x_0)$.*

Proof Consider the commutative diagram

$$
\begin{array}{ccccccc}
0 & \longrightarrow & \widetilde{K}^p(X, x_0) & \longrightarrow & K^p(X) & \stackrel{j^*}{\longrightarrow} & K^p(x_0) \\
& & \Big\| & & \Big\downarrow = & & \Big\downarrow = \\
0 & \longrightarrow & K^p(X, x_0) & \longrightarrow & K^p(X) & \stackrel{j^*}{\longrightarrow} & K^p(x_0).
\end{array}
$$

The top row is exact by definition, and the exactness of the bottom row is a consequence of Theorems 2.2.10 and 2.3.15. Proposition 1.8.9 then gives us an isomorphism $\widetilde{K}^p(X, x_0) \cong K^p(X, x_0)$ whose naturality follows easily from the commutativity of the diagram. □

Corollary 2.5.4 *For every pointed space (X, x_0), there exist isomorphisms*

$$
K^0(X) \cong \widetilde{K}^0(X, x_0) \oplus K^0(x_0) \cong \widetilde{K}^0(X, x_0) \oplus \mathbb{Z}
$$
$$
K^{-1}(X) \cong \widetilde{K}^{-1}(X, x_0) \oplus K^{-1}(x_0) \cong \widetilde{K}^{-1}(X, x_0).
$$

Proof These isomorphisms are immediate consequences of Proposition 2.5.3 and Corollaries 2.2.11 and 2.3.16. □

Corollary 2.5.5 *For $p = 0, -1$, the assignment $(X, x_0) \mapsto \widetilde{K}^p(X, x_0)$ determines a contravariant functor from pointed spaces to abelian groups.*

Proof The category of pointed spaces is a subcategory of the category of compact pairs, so the corollary follows from Propositions 2.2.7, 2.3.13, and 2.5.3. □

Proposition 2.5.6 *Let (X, x_0) be a pointed space. Then every element of $\widetilde{K}^0(X, x_0)$ can be written in the form $[\mathsf{E}] - [I_k]$, where E is an idempotent in $\mathrm{M}(n, C(X))$ with the property that $\mathsf{E}(x_0) = \mathrm{diag}(I_n, 0_{n-k})$.*

Proof If $[\mathsf{E}] - [I_k]$ is in $\widetilde{K}^0(X, x_0) \subseteq K^0(X)$, then the rank of the vector space $\mathsf{E}(x_0)$ is k. We use elementary linear algebra to produce a matrix

U in $\mathrm{GL}(n,\mathbb{C})$ such that $\mathsf{U E U}^{-1}(x_0) = \mathrm{diag}(I_k, 0_{n-k})$; replacing E with $\mathsf{U E U}^{-1}$ yields the desired result. $\square$

Theorem 2.5.7 *Let (X, A) be a compact pair, fix a point a_0 in A, and let $j : A \longrightarrow X$ be the inclusion map. Then the sequence*

$$\mathrm{K}^{-1}(X, A) \xrightarrow{\mu_1^*} \widetilde{\mathrm{K}}^{-1}(X, a_0) \xrightarrow{j^*} \widetilde{\mathrm{K}}^{-1}(A, a_0) \xrightarrow{\partial}$$

$$\mathrm{K}^0(X, A) \xrightarrow{\mu_1^*} \widetilde{\mathrm{K}}^0(X, a_0) \xrightarrow{j^*} \widetilde{\mathrm{K}}^0(A, a_0)$$

is exact.

Proof Exactness at $\widetilde{\mathrm{K}}^{-1}(X, a_0)$ and $\widetilde{\mathrm{K}}^{-1}(A, a_0)$ follows immediately from Theorem 2.4.1 and Corollary 2.5.4. In addition, Corollary 2.5.4 states that $\mathrm{K}^0(X) \cong \widetilde{\mathrm{K}}^0(X, a_0) \oplus \mathrm{K}^0(a_0)$ and $\mathrm{K}^0(A) \cong \widetilde{\mathrm{K}}^0(A, a_0) \oplus \mathrm{K}^0(a_0)$, and Theorem 2.4.1 gives us exactness at $\mathrm{K}^0(X, A)$ and $\widetilde{\mathrm{K}}^0(X, a_0)$. $\square$

2.6 *K*-theory of locally compact topological spaces

In this section we extend the definitions of $\mathrm{K}^0(X)$ and $\mathrm{K}^{-1}(X)$ to *locally compact* topological spaces. The *K*-theory of such spaces is not only useful in applications, but it will also show us how to define groups $\mathrm{K}^{-n}(X)$ for every natural number n. In fact, we will eventually allow n to be any integer, but this development must wait until later in the chapter.

Definition 2.6.1 *A Hausdorff topological space X is* locally compact *if every point of X has an open neighborhood whose closure in X is compact. Let ∞_X be a point not in X. The* one-point compactification X^+ *of X is the set $X \cup \{\infty_X\}$, equipped with the following topology: a subset of X^+ is open if it is open in X or if it can be written $X^+\backslash C$ for some compact subset of X.*

If the topological space X is clear from context, we usually write just ∞ instead of ∞_X.

Every compact Hausdorf space X is locally compact Hausdorff; in this case X^+ is the disjoint union of X and $\{\infty\}$. Open or closed subspaces of a locally compact Hausdorff space are also locally compact. However, not every subspace of a locally compact Hausdorff space is locally compact; for example, the rational numbers $\mathbb{Q}$ do not form a locally compact subspace of $\mathbb{R}$.

Definition 2.6.2 *Let X and Y be locally compact Hausdorff spaces and suppose $\phi : X \longrightarrow Y$ is continuous. We say ϕ is* continuous at infinity *if ϕ is the restriction of a continuous function $\phi^+ : X^+ \longrightarrow Y^+$.*

If X and Y are compact Hausdorff, then every continuous function ϕ from X to Y is continuous at infinity; the points ∞_x and ∞_Y are isolated from X and Y respectively, so we can simply decree that $\phi^+(\infty_X) = \infty_Y$.

Definition 2.6.3 *A continuous function $\phi : X \longrightarrow Y$ between topological spaces is* proper *if $\phi^{-1}(D)$ is a compact subset of X for every compact subset D of Y.*

Proposition 2.6.4 *Suppose $\phi : X \longrightarrow Y$ is a proper map between locally compact Hausdorff spaces. Then ϕ is continuous at infinity.*

Proof Extend ϕ to a function $\phi^+ : X^+ \longrightarrow Y^+$ by setting $\phi^+(\infty_X) = \infty_Y$. To see that ϕ^+ is continuous, first suppose that V is open in Y. Then $(\phi^+)^{-1}(V) = \phi^{-1}(V)$, which is open in X and hence open in X^+. Now suppose D is a compact subset of Y. Then

$$(\phi^+)^{-1}(Y^+\backslash D) = (\phi^+)^{-1}(Y^+)\backslash(\phi^+)^{-1}(D) = X^+\backslash(\phi^+)^{-1}(D),$$

which is open in X^+. Therefore ϕ is continuous at infinity. $\square$

Not every function continuous at infinity is proper (Exercise 2.4).

Corollary 2.6.5 *Suppose that X is locally compact Hausdorff and that A is a closed subspace of X. Then the inclusion $j : A \longrightarrow X$ is continuous at infinity.*

Proof For every compact subset C in X, the inverse image $j^{-1}(C) = A \cap C$ is compact in A because A is closed in X. Thus j is proper. $\square$

In light of Definition 2.6.2 and Corollary 2.6.5, we may identify ∞_A with ∞_X and consider A^+ as a subspace of X^+ whenever X is locally compact and A is closed in X.

Because the collection of functions continuous at infinity is closed under compositions, the class of locally compact Hausdorff spaces with functions continuous at infinity as morphisms forms a category.

Proposition 2.6.6 *The assignment $X \mapsto (X^+, \infty)$ defines a covariant functor from the category of locally compact Hausdorff spaces to the category of pointed spaces.*

Proof If X and Y are locally compact Hausdorff and $\phi : X \longrightarrow Y$ is continuous at infinity, then ϕ extends to a continuous function ϕ^+ from X^+ to Y^+, and $\phi^+(\infty_X) = \infty_Y$. Obviously the identity map on X extends to the identity map on X^+. Finally, if Z is locally compact Hausdorff and $\psi : Y \longrightarrow Z$ is continuous at infinity, then $(\psi\phi)^+ = \psi^+\phi^+$. $\qquad\square$

Definition 2.6.7 *For $p = 0, -1$ and X locally compact Hausdorff, we define* $\mathrm{K}^\mathrm{p}(X) = \widetilde{\mathrm{K}}^\mathrm{p}(X^+)$.

If X is actually compact, then Examples 2.1.3 and 2.3.5 imply that $\mathrm{K}^\mathrm{p}(X^+) \cong \mathrm{K}^\mathrm{p}(X) \oplus \mathrm{K}^\mathrm{p}(\{\infty\})$, whence $\widetilde{\mathrm{K}}^\mathrm{p}(X^+)$ isomorphic to $\mathrm{K}^\mathrm{p}(X)$. Therefore our new definitions of $\mathrm{K}^0(X)$ and $\mathrm{K}^{-1}(X)$ agree with the old ones when X is compact.

Proposition 2.6.8 *For every compact pair (X, A), there exist natural isomorphisms* $\mathrm{K}^\mathrm{p}(X, A) \cong \mathrm{K}^\mathrm{p}(X\backslash A)$ *for $p = 0, -1$.*

Proof By Theorems 2.2.14 and 2.3.17, we have natural isomorphisms

$$\mathrm{K}^\mathrm{p}(X, A) \cong \mathrm{K}^\mathrm{p}\big((X\backslash A)^+, \infty\big) = \widetilde{\mathrm{K}}^\mathrm{p}\big((X\backslash A)^+\big) = \mathrm{K}^\mathrm{p}(X\backslash A).$$

$\qquad\square$

Proposition 2.6.9 *Let X be a locally compact Hausdorff space. Then every element of $\mathrm{K}^0(X)$ can be written in the form $[\mathsf{E}] - [I_k]$, where E is an idempotent in $\mathrm{M}(n, C(X^+))$ whose value at ∞ is I_k.*

Proof Follows immediately from Proposition 2.5.6 and Definition 2.6.7. $\qquad\square$

Note that $\mathrm{K}^{-1}(X) \cong \mathrm{K}^{-1}(X^+)$ by Proposition 2.5.4, so the same matrices that represent elements of $\mathrm{K}^{-1}(X^+)$ also represent elements of $\mathrm{K}^{-1}(X)$.

Proposition 2.6.10 *Both K^0 and K^{-1} are contravariant functors from the category of locally compact Hausdorff spaces to the category of abelian groups.*

Proof Follows immediately from Corollary 2.5.5 and Proposition 2.6.6. $\qquad\square$

Theorem 2.6.11 *Suppose that A is a closed subspace of a locally compact Hausdorff space X and let $j : A \longrightarrow X$ be the inclusion map. Then for $p = 0, -1$, there exists an exact sequence*

$$\mathrm{K}^{\mathrm{p}}(X \backslash A) \longrightarrow \mathrm{K}^{\mathrm{p}}(X) \xrightarrow{\ j^*\ } \mathrm{K}^{\mathrm{p}}(A).$$

Furthermore, if $\psi : X \longrightarrow A$ is a function continuous at infinity and has the property that ψj is the identity map on A, then there exists a split exact sequence

$$0 \longrightarrow \mathrm{K}^{\mathrm{p}}(X \backslash A) \longrightarrow \mathrm{K}^{\mathrm{p}}(X) \underset{\psi^*}{\overset{j^*}{\rightleftarrows}} \mathrm{K}^{\mathrm{p}}(A) \longrightarrow 0,$$

whence $\mathrm{K}^{\mathrm{p}}(X)$ is isomorphic to $\mathrm{K}^{\mathrm{p}}(A) \oplus \mathrm{K}^{\mathrm{p}}(X \backslash A)$.

Proof We have the exact sequence

$$\mathrm{K}^{\mathrm{p}}(X^+, A^+) \xrightarrow{\ \mu_1^*\ } \mathrm{K}^{\mathrm{p}}(X^+) \xrightarrow{\ j^*\ } \mathrm{K}^{\mathrm{p}}(A^+)$$

from Theorems 2.2.9 and 2.3.14. Because $X^+ \backslash A^+ = X \backslash A$, Proposition 2.6.8 gives us the first exact sequence in the statement of the theorem. In a similar fashion, the split exact sequence is a consequence of Theorems 2.2.10 and 2.3.15 applied to the compact pair (X^+, A^+). $\qquad\square$

The proof of Theorem 2.6.11 brings up an issue that we mention in passing. Suppose that X is locally compact Hausdorff and that U is an open subset of X. Then the inclusion of U into X induces homomorphisms from $\mathrm{K}^{\mathrm{p}}(U) = \mathrm{K}^{\mathrm{p}}(X \backslash (X \backslash U))$ into $\mathrm{K}^{\mathrm{p}}(X)$ for $p = 0, -1$. Thus the inclusion of an *open* subspace into a locally compact Hausdorff space induces a group homomorphism that goes in the opposite direction from the homomorphism induced by the inclusion of a *closed* subspace into a locally compact Hausdorff space.

Theorem 2.6.12 *If A is a closed subspace of a locally compact Hausdorff space X and $j : A \longrightarrow X$ denotes the inclusion map, then there exists an exact sequence*

$$\mathrm{K}^{-1}(X \backslash A) \longrightarrow \mathrm{K}^{-1}(X) \xrightarrow{\ j^*\ } \mathrm{K}^{-1}(A) \xrightarrow{\ \partial\ }$$

$$\mathrm{K}^0(X \backslash A) \longrightarrow \mathrm{K}^0(X) \xrightarrow{\ j^*\ } \mathrm{K}^0(A).$$

Furthermore, the homomorphism ∂ determines a natural transformation from K^{-1} to K^0.

Proof Theorem 2.6.11 gives us exactness at $K^0(X)$ and $K^{-1}(X)$. To show the sequence is exact at $K^{-1}(A)$ and $K^0(X \backslash A)$, consider the diagram

$$
\begin{array}{ccccccc}
K^{-1}(X) & \xrightarrow{j^*} & K^{-1}(A) & \dashrightarrow & K^0(X \backslash A) & \longrightarrow & K^0(X) \\
\downarrow & & \downarrow & & \downarrow & & \downarrow \\
\widetilde{K}^{-1}(X^+) & \xrightarrow{j^*} & \widetilde{K}^{-1}(A^+) & \xrightarrow{\partial} & K^0(X^+, A^+) & \xrightarrow{j^*} & \widetilde{K}^0(A^+).
\end{array}
$$

The vertical maps are isomorphisms by Corollary 2.5.4, and thus we can uniquely define a homomorphism ∂ from $K^{-1}(A)$ to $K^0(X \backslash A)$ that makes the middle square commute. To describe the connecting homomorphism explicitly, take S in $GL(n, C(A^+))$ and use Corollary 1.3.17 to find T in $GL(2n, C(X^+))$ whose restriction to A^+ is $\mathrm{diag}(S, S^{-1})$. The matrix $T \, \mathrm{diag}(I_n, 0_n) T^{-1}$ has the constant value $\mathrm{diag}(I_n, 0_n)$ when restricted to A, and hence can be considered an element of $M(2n, C((X \backslash A)^+))$. Then

$$
\partial[S] = [T \, \mathrm{diag}(I_n, 0_n) T^{-1}] - [I_n].
$$

Theorem 2.5.7 states that the bottom row of the diagram is exact. This fact, combined with the commutativity of the two outer squares, shows that the top row is exact. Finally, the homomorphism ∂ is natural because it is a composition of natural maps. □

The extension of K-theory to locally compact Hausdorff spaces gives us a perhaps unexpected relationship between K^0 and K^{-1}.

Theorem 2.6.13 *For every locally compact Hausdorff space X, there is a natural isomorphism* $K^{-1}(X) \cong K^0(X \times \mathbb{R})$.

Proof For each $0 \leq t \leq 1$, define ϕ_t from $(X \times (0, 1])^+$ to itself by the formulas

$$
\begin{aligned}
\phi_t(x, s) &= (x, (1 - t)s) \quad \text{for } (x, s) \text{ in } X \times (0, 1] \\
\phi_t(\infty) &= \infty.
\end{aligned}
$$

Then $\{\phi_t\}$ is a homotopy from the identity map on $(X \times (0, 1])^+$ to the constant function whose range is $\{\infty\}$, and thus $(X \times (0, 1])^+$ is contractible. Corollaries 2.1.8 and 2.3.10 and Definition 2.6.7 imply that $K^p(X \times (0, 1])$ is trivial for $p = 0, -1$.

Apply Theorem 2.6.12 to the inclusion of $X \times \{1\}$ into $X \times (0, 1]$ to obtain an isomorphism $K^{-1}(X \times \{1\}) \cong K^0(X \times (0, 1))$; the homeomorphisms $X \times \{1\} \cong X$ and $(0, 1) \cong \mathbb{R}$ then yield the desired isomorphism.

Moreover, this isomorphism is a composition of natural maps and is therefore a natural transformation. $\square$

Theorem 2.6.13 allows us to extend the exact sequence in Theorem 2.4.1 infinitely to the left.

Definition 2.6.14 *We define* $K^{-n}(X) = K^0(X \times \mathbb{R}^n)$ *for every natural number n and locally compact Hausdorff space X.*

Proposition 2.6.15 *For each natural number n, we have a contravariant functor* K^{-n} *from locally compact Hausdorff spaces to abelian groups.*

Proof Follows directly from Proposition 2.6.10. $\square$

Theorem 2.6.16 *Suppose that A is a closed subspace of a locally compact Hausdorff space X and let* $j : A \longrightarrow X$ *be the inclusion map. Then for each natural number n, there exists a natural homomorphism* $\partial : K^{-n-1}(X) \longrightarrow K^{-n}(X \backslash A)$ *that makes the infinite sequence*

$$\cdots \longrightarrow K^{-2}(X) \xrightarrow{\;j^*\;} K^{-2}(A) \xrightarrow{\;\partial\;}$$

$$K^{-1}(X \backslash A) \longrightarrow K^{-1}(X) \xrightarrow{\;j^*\;} K^{-1}(A) \xrightarrow{\;\partial\;}$$

$$K^0(X \backslash A) \longrightarrow K^0(X) \xrightarrow{\;j^*\;} K^0(A)$$

exact.

Proof For each natural number $n > 1$, apply Theorem 2.6.12 to the inclusion of $A \times \mathbb{R}^n$ into $X \times \mathbb{R}^n$. $\square$

2.7 Bott periodicity

Thanks to the results of the previous section, we now have an infinite collection of contravariant K-functors from locally compact Hausdorff spaces to abelian groups. However, as we shall see in this section, our original functors K^0 and K^{-1} are the only ones up to isomorphism. This surprising result is the central theorem in K-theory, and is called the *Bott periodicity theorem.*

Theorem (Bott periodicity) *For every locally compact Hausdorff* *space X, there exists a natural isomorphism*

$$\beta : \mathrm{K}^0(X) \longrightarrow \mathrm{K}^{-2}(X).$$

The Bott periodicity theorem is quite difficult to prove and involves many steps, so we begin with an overview of the proof. We will first establish the theorem for compact Hausdorff spaces; the extension to locally compact Hausdorff spaces will then follow easily from homological algebra. Second, instead of mapping directly into $\mathrm{K}^{-2}(X) = \mathrm{K}^{-1}(X \times \mathbb{R})$, we will replace the noncompact Hausdorff space $X \times \mathbb{R}$ with its one-point compactification $(X \times \mathbb{R})^+$; these two topological spaces have isomorphic K^{-1} groups due to the fact that K^{-1} of a point is trivial.

The difficult part of the proof is showing that β is a bijection. We will accomplish this by showing that elements of $\mathrm{K}^{-1}\big((X \times \mathbb{R})^+\big)$ can be represented by invertible matrices with particularly simple entries. We first use Fourier series to show that every element of $\mathrm{K}^{-1}\big((X \times \mathbb{R})^+\big)$ can be represented by a "polynomial" matrix, and we will reduce the degree of the polynomial by enlarging the size of the matrix. Once we have an element of $\mathrm{K}^{-1}\big((X \times \mathbb{R})^+\big)$ represented by a linear polynomial, we will show that this polynomial can be written in a specific form that will allow us to see why β is bijective. These reductions come at a price: we will be working with extremely large matrices!

We begin by simplifying some of our notation.

Definition 2.7.1 *Let S^1 denote the unit circle in $\mathbb{C}$. For every compact Hausdorff space X, define $\mathcal{S}'X$ to be the topological space obtained by taking the product $X \times S^1$ and collapsing $X \times \{1\}$ to a point.*

Note that $\mathcal{S}'X$ is homeomorphic to $(X \times \mathbb{R})^+$.

We will use Lemma 2.2.13 to write matrices over $\mathcal{S}'X$ as matrices over $X \times S^1$ that are constant on $X \times \{1\}$. We will use the variable z for points on S^1, and to simplify notation in many of the proofs that follow, we will usually suppress the dependence on X when working with matrices over $X \times S^1$ and $\mathcal{S}'X$.

The topological space $\mathcal{S}'X$ is similar to, but not the same as, the *suspension* of X (Exercise 2.10).

Proposition 2.7.2 *Let X be a compact Hausdorff space. For each natural number n and idempotent E in $\mathrm{M}(n, C(X))$ set*

$$\beta[\mathsf{E}] = [I_n - \mathsf{E} + \mathsf{E}z].$$

Then β is a natural group homomorphism from $\mathrm{K}^0(X)$ to $\mathrm{K}^{-1}(S'X)$.

Proof The matrix $I_n - \mathsf{E} + \mathsf{E}z$ is constant when $z = 1$ and is invertible with inverse $I_n - \mathsf{E} + \mathsf{E}z^{-1}$. Therefore $I_n - \mathsf{E} + \mathsf{E}z$ determines an element of $\mathrm{K}^{-1}(S'X)$. Because

$$\beta[(\mathrm{diag}(\mathsf{E}, 0))] = [I_{n+1} - \mathrm{diag}(\mathsf{E}, 0) + \mathrm{diag}(\mathsf{E}, 0)z]$$
$$= [\mathrm{diag}(I_n - \mathsf{E} + \mathsf{E}z, 1)]$$
$$= [I_n - \mathsf{E} + \mathsf{E}z],$$

we see that $\beta[\mathsf{E}]$ does not depend upon the size of matrix used to represent E. Furthermore, if $\{\mathsf{E}_t\}$ is a homotopy of idempotents in $\mathrm{M}(n, C(X))$, then $\{I_n - \mathsf{E}_t + \mathsf{E}_t z\}$ is a homotopy of invertibles, whence $\beta[\mathsf{E}]$ is well defined by Proposition 1.4.9.

To see that β is a group homomorphism, take idempotents E and F in $\mathrm{M}(n, C(X))$ and $\mathrm{M}(m, C(X))$ respectively. Then Corollary 2.3.7 gives us

$$\beta([\mathsf{E}] + [\mathsf{F}]) = \beta[\mathrm{diag}(\mathsf{E}, \mathsf{F})]$$
$$= [I_{n+m} - \mathrm{diag}(\mathsf{E}, \mathsf{F}) + \mathrm{diag}(\mathsf{E}, \mathsf{F})z]$$
$$= [I_n - \mathsf{E} + \mathsf{E}z][I_m - \mathsf{F} + \mathsf{F}z]$$
$$= \beta[\mathsf{E}]\beta[\mathsf{F}].$$

Thus β is a monoid homomorphism, and if we set

$$\beta([\mathsf{E}] - [\mathsf{E}']) = \beta[\mathsf{E}](\beta[\mathsf{E}'])^{-1}$$

for each element $[\mathsf{E}] - [\mathsf{E}']$ in $\mathrm{K}^0(C(X))$, Theorem 1.6.7 implies that β is a group homomorphism.

Finally, suppose Y is a compact Hausdorff space and that $\phi : Y \longrightarrow X$ is continuous. Then

$$\phi^* \beta[\mathsf{E}] = \beta\phi^*[\mathsf{E}] = [I_n - \phi^*\mathsf{E} + (\phi^*\mathsf{E})z]$$

for all natural numbers n and idempotents E in $\mathrm{M}(n, C(X))$, whence β is natural. $\square$

Representatives of elements of $\mathrm{K}^{-1}(S'X)$ can be thought of as loops of invertible matrices whose entries are in $C(X)$; i.e., by continuous functions from S^1 to $\mathrm{GL}(C(X))$. Such functions can be approximated by matrix-valued Fourier series.

Definition 2.7.3 *Let X be compact Hausdorff and let n be a natural number. For each* T *in* $\mathrm{M}(n, C(X \times S^1))$ *and each integer k, let*

$$\widehat{\mathsf{T}}_k = \frac{1}{2\pi} \int_{-\pi}^{\pi} \mathsf{T}(\mu) e^{-ik\mu} \, d\mu$$

in $\mathrm{M}(n, C(X))$ *denote the kth Fourier coefficient of* T.

For each natural number m, define the mth Cesàro mean of T *to be*

$$\mathsf{C}_m = \frac{1}{m+1} \sum_{j=0}^{m} \sum_{k=-j}^{j} \widehat{\mathsf{T}}_k e^{ik\theta}.$$

The point of working with Cesàro means is that while the partial sums of the Fourier series of T may not converge uniformly to T, the Cesàro means do.

Theorem 2.7.4 (Fejér) *Let X be a compact Hausdorff space, let n be a natural number, and take* T *in* $\mathrm{M}(n, C(X \times S^1))$. *Then the Cesàro means of* T *converge uniformly to* T; *i.e.,* $\lim_{m \to \infty} \|\mathsf{C}_m - \mathsf{T}\|_\infty = 0$.

Proof We begin by writing the Cesàro means of T in an alternate form. Using the definition of C_m, a change of variable, and the periodicity of T, we obtain

$$\mathsf{C}_m = \frac{1}{m+1} \sum_{j=0}^{m} \sum_{k=-j}^{j} \widehat{\mathsf{T}}_k e^{ik\theta}$$

$$= \frac{1}{m+1} \sum_{j=0}^{m} \sum_{k=-j}^{j} \frac{1}{2\pi} \int_{-\pi}^{\pi} \mathsf{T}(\mu) e^{ik(\theta-\mu)} \, d\mu$$

$$= \frac{1}{2\pi(m+1)} \sum_{j=0}^{m} \sum_{k=-j}^{j} \int_{\theta-\pi}^{\theta+\pi} \mathsf{T}(\theta - \alpha) e^{ik\alpha} \, d\alpha$$

$$= \frac{1}{2\pi(m+1)} \sum_{j=0}^{m} \sum_{k=-j}^{j} \int_{-\pi}^{\pi} \mathsf{T}(\theta - \alpha) e^{ik\alpha} \, d\alpha$$

$$= \frac{1}{2\pi} \int_{-\pi}^{\pi} \mathsf{T}(\theta - \alpha) \left(\sum_{j=0}^{m} \sum_{k=-j}^{j} \frac{e^{ik\alpha}}{m+1} \right) \, d\alpha.$$

The sum in parentheses is denoted $K_m(\alpha)$ and is called the *Fejér kernel*.

An involved, but elementary, calculation yields

$$K_m(\alpha) = \begin{cases} \dfrac{1}{m+1}\dfrac{\sin^2(\alpha(m+1)/2)}{\sin^2(\alpha/2)} & \text{if } \sin(\alpha/2) \neq 0 \\ m+1 & \text{if } \sin(\alpha/2) = 0. \end{cases}$$

Comparing the two expressions for the Fejér kernel, we see that it satisfies

$$K_m(\alpha) \geq 0 \tag{1}$$

$$K_m(-\alpha) = K_m(\alpha) \tag{2}$$

$$0 \leq K_m(\alpha) \leq m+1 \tag{3}$$

$$\frac{1}{2\pi}\int_{-\pi}^{\pi} K_m(\alpha)\,d\alpha = 1 \tag{4}$$

for all α and m. The graph of $y = \sin\theta$ is concave down on the interval $(0, \pi/2)$ and thus lies above the line segment from $(0,0)$ to $(\pi/2, 1)$; in other words, we have the inequality $\sin\theta > 2\theta/\pi$ for all $0 < \theta < \pi/2$. This gives us the additional property

$$K_m(\alpha) \leq \frac{1}{m+1}\frac{1}{\sin^2(\alpha/2)} \leq \frac{1}{m+1}\frac{\pi^2}{\alpha^2}, \quad 0 < |\alpha| < \pi. \tag{5}$$

Property (4) implies that

$$\mathsf{T}(\theta) = \frac{1}{2\pi}\int_{-\pi}^{\pi} \mathsf{T}(\theta)K_m(\alpha)\,d\alpha,$$

and thus

$$\|\mathsf{C}_m(\theta) - \mathsf{T}(\theta)\|_\infty = \left\|\frac{1}{2\pi}\int_{-\pi}^{\pi}(\mathsf{T}(\theta-\alpha) - \mathsf{T}(\theta))K_m(\alpha)\,d\alpha\right\|_\infty$$

$$\leq \frac{1}{2\pi}\int_{-\pi}^{\pi}\|\mathsf{T}(\theta-\alpha) - \mathsf{T}(\theta)\|_\infty K_m(\alpha)\,d\alpha;$$

the inequality follows from property (1).

Fix $\epsilon > 0$. Because $X \times S^1$ is compact, the function T is uniformly continuous, and so there exists a $\delta > 0$ such that $|\alpha| < \delta$ implies

$$\|\mathsf{T}(\theta-\alpha) - \mathsf{T}(\theta)\|_\infty < \frac{\epsilon}{2}.$$

Then

$$\frac{1}{2\pi} \int_{-\delta}^{\delta} \|T(\theta - \alpha) - T(\theta)\|_\infty K_m(\alpha) \, d\alpha$$

$$< \frac{\epsilon}{4\pi} \int_{-\delta}^{\delta} K_m(\alpha) \, d\alpha \leq \frac{\epsilon}{4\pi} \int_{-\pi}^{\pi} K_m(\alpha) \, d\alpha = \frac{\epsilon}{2}.$$

On the other hand, property (5) gives us

$$\frac{1}{2\pi} \int_{\delta}^{\pi} \|T(\theta - \alpha) - T(\theta)\|_\infty K_m(\alpha) \, d\alpha \leq \frac{1}{\pi} \|T\|_\infty \int_{\delta}^{\pi} K_m(\alpha) \, d\alpha$$

$$< \frac{1}{\pi} \|T\|_\infty \frac{\pi^2}{m+1} \int_{\delta}^{\pi} \frac{1}{\alpha^2} \, d\alpha$$

$$= \frac{\pi - \delta}{\delta(m+1)} \|T\|_\infty.$$

Similarly, from property (2) we know

$$\frac{1}{2\pi} \int_{-\pi}^{-\delta} \|T(\theta - \alpha) - T(\theta)\|_\infty K_m(\alpha) \, d\alpha < \frac{\pi - \delta}{\delta(m+1)} \|T\|_\infty,$$

and thus

$$\|C_m(\theta) - T(\theta)\|_\infty < \frac{\epsilon}{2} + \frac{2(\pi - \delta)}{\delta(m+1)} \|T\|_\infty.$$

Therefore for m sufficiently large we obtain $\|C_m - T\|_\infty < \epsilon$, as desired.

$\square$

Definition 2.7.5 *Let X be a compact Hausdorff space and let n be a natural number. An element T of $M(n, C(X \times S^1))$ is called a Laurent polynomial if there exist a natural number m and elements A_k in $M(n, C(X))$ for $-m \leq k \leq m$ such that*

$$T(z) = \sum_{k=-m}^{m} A_k z^k.$$

If either A_m or A_{-m} is not identically zero, we say T has degree m. If $T(z) = \sum_{k=0}^{m} A_k z^k$, then T is a polynomial.

Definition 2.7.6 *Let X be compact Hausdorff and let m and n be natural numbers. Define*

$$\mathcal{L}^m(n, C(\mathcal{S}'X)) = \{L \in GL(n, C(\mathcal{S}'X)) : L \text{ is a Laurent polynomial}$$
$$\text{of degree at most } m \text{ such that } L(1) = I_n\}$$

$$\mathcal{P}^m(n, C(\mathcal{S}'X)) = \{\mathsf{P} \in \mathrm{GL}(n, C(\mathcal{S}'X)) : \mathsf{P} \text{ is a polynomial}$$
$$\text{of degree at most } m \text{ such that } \mathsf{P}(1) = I_n\}$$

$$\mathcal{I}(n, C(\mathcal{S}'X)) = \{I_n - \mathsf{E} + \mathsf{E}z : \mathsf{E} \text{ is an idempotent in } \mathrm{M}(n, C(X))\}.$$

Lemma 2.7.7 *Let X be a compact Hausdorff space, let n be a natural number, and suppose that T is in $\mathrm{GL}(n, C(\mathcal{S}'X))$. Then for every $\epsilon > 0$, there exists a natural number m and a Laurent polynomial L of degree at most m such that $\|\mathsf{T} - \mathsf{L}\|_\infty < \epsilon$. In addition, if $\mathsf{T}(1) = I_n$, then L can be chosen to be in $\mathcal{L}^m(n, C(\mathcal{S}'X))$.*

Proof Fix $\epsilon > 0$ and choose $\delta \leq 1/\|\mathsf{T}^{-1}\|_\infty$. By Theorem 2.7.4, there exists a natural number M with the property that $\|\mathsf{T} - \mathsf{C}_m\|_\infty < \delta$ for all $m \geq M$, and Proposition 1.3.5 states that C_m is invertible for all such m.

Now suppose $\mathsf{T}(1) = I_n$ and define $\mathsf{L}_m(z) = \mathsf{C}_m(z)\mathsf{C}_m(1)^{-1}\mathsf{T}(1)$ for $m \geq M$. Then L_m is invertible and

$$\begin{aligned}
\|\mathsf{T} - \mathsf{L}_m\|_\infty &= \left\|\mathsf{T} - \mathsf{C}_m\mathsf{C}_m(1)^{-1}\mathsf{T}(1)\right\|_\infty \\
&\leq \|\mathsf{T} - \mathsf{C}_m\|_\infty + \left\|\mathsf{C}_m - \mathsf{C}_m\mathsf{C}_m(1)^{-1}\mathsf{T}(1)\right\|_\infty \\
&\leq \|\mathsf{T} - \mathsf{C}_m\|_\infty + \|\mathsf{C}_m\|_\infty \left\|I_n - \mathsf{C}_m(1)^{-1}\mathsf{T}(1)\right\|_\infty \\
&\leq \delta + \left(\|\mathsf{T}\|_\infty + \delta\right)\left\|I_n - \mathsf{C}_m(1)^{-1}\mathsf{T}(1)\right\|_\infty.
\end{aligned}$$

We may therefore choose δ sufficiently small and m sufficiently large so that $\|\mathsf{T} - \mathsf{L}_m\|_\infty < \epsilon$. Setting $\mathsf{L} = \mathsf{L}_m$ gives us the desired element of $\mathcal{L}^m(n, C(\mathcal{S}'X))$. □

Lemma 2.7.8 *Let X be compact Hausdorff, let m and n be natural numbers, and suppose that L_0 and L_1 in $\mathcal{L}^m(n, C(\mathcal{S}'X))$ are homotopic in $\mathrm{GL}(n, C(\mathcal{S}'X))$. Then for some natural number $M \geq m$, there exists a homotopy in $\mathcal{L}^M(n, C(\mathcal{S}'X))$ from L_0 to L_1.*

Proof Choose a homotopy in $\mathrm{GL}(n, C(\mathcal{S}'X))$ from L_0 to L_1; we write this homotopy as $\{\mathsf{H}(t)\}$ instead of the more usual $\{\mathsf{H}_t\}$. Set

$$R = \sup\{\sqrt{2} + \|\mathsf{H}(t)^{-1}\|_\infty : 0 \leq t \leq 1\}$$

and choose points $0 = t_0 < t_1 < t_2 < \cdots < t_K = 1$ so that

$$\|\mathsf{H}(t_k) - \mathsf{H}(t_{k-1})\|_\infty < \frac{1}{3R^3}$$

for all $0 < k \leq K$. For each $0 < k < K$, we use Lemma 2.7.7 to choose an integer $m(k) \geq m$ and an element $\mathsf{L}(t_k)$ in $\mathcal{L}^{m(k)}(n, C(\mathcal{S}'X))$ with the property that

$$\|\mathsf{L}(t_k) - \mathsf{H}(t_k)\|_\infty < \frac{1}{3R^3}.$$

Each $\mathsf{L}(t_k)$ is invertible by Proposition 1.3.5 and

$$\|\mathsf{L}(t_k) - \mathsf{L}(t_{k-1})\|_\infty \leq \|\mathsf{L}(t_k) - \mathsf{H}(t_k)\|_\infty +$$
$$\|\mathsf{H}(t_k) - \mathsf{H}(t_{k-1})\|_\infty + \|\mathsf{H}(t_{k-1}) - \mathsf{L}(t_{k-1})\|_\infty < \frac{1}{R^3}$$

for all $0 < k \leq K$. Next, Lemma 1.3.7 gives us

$$\left\|\mathsf{L}(t_k)^{-1}\right\|_\infty \leq \left\|\mathsf{L}(t_k)^{-1} + \mathsf{H}(t_k)^{-1}\right\|_\infty + \left\|\mathsf{H}(t_k)^{-1}\right\|_\infty$$
$$< 2R^2 \cdot \frac{1}{R} + R$$
$$= \frac{2 + R^2}{R}.$$

Because $R > \sqrt{2}$, we obtain the string of inequalities

$$\|\mathsf{L}(t_k) - \mathsf{L}(t_{k-1})\|_\infty < \frac{1}{R^3} < \frac{R}{2 + R^2} < \frac{1}{\|\mathsf{L}(t_k)^{-1}\|_\infty},$$

and Proposition 1.3.5 implies that for each $0 < k \leq K$, the straight line homotopy from $\mathsf{L}(t_{k-1})$ to $\mathsf{L}(t_k)$ consists of invertible elements. Concatenate these straight line homotopies to obtain a homotopy $\{\mathsf{L}_t\}$ of invertible Laurent polynomials of degree at most $M = \max\{m(k) : 0 \leq k \leq K\}$ from L_0 to L_1. Finally, if we define $\widetilde{\mathsf{L}}_t(z) = \mathsf{L}_t(z)\mathsf{L}_t^{-1}(1)$, then $\{\widetilde{\mathsf{L}}_t\}$ is a homotopy in $\mathcal{L}^M(n, C(\mathcal{S}'X))$ from L_0 to L_1. $\qquad\square$

Proposition 2.7.9 *Let X be compact Hausdorff, let n be a natural number, and suppose T is an element of $\mathrm{GL}(n, C(\mathcal{S}'X))$.*

(i) *There exist a natural number m and a Laurent polynomial L in $\mathcal{L}^m(n, C(\mathcal{S}'X))$ such that $[\mathsf{L}] = [\mathsf{T}]$ in $\mathrm{K}^{-1}(\mathcal{S}'X)$.*

(ii) *For some natural number M, there exist polynomials P and Q in $\mathcal{P}^M(n, C(\mathcal{S}'X))$ such that $[\mathsf{T}] = [\mathsf{P}][\mathsf{Q}]^{-1}$ in $\mathrm{K}^{-1}(\mathcal{S}'X)$.*

Proof Thanks to Proposition 1.3.5 and Lemma 2.7.7, we may assume that T is a Laurent polynomial. Define $\mathsf{L}(z) = \mathsf{T}(z)\mathsf{T}(1)^{-1}$. Then L is in $\mathcal{L}^m(n, C(\mathcal{S}'X))$ for some natural number m. The matrix $\mathsf{T}(1)^{-1}$ is a constant matrix, and Proposition 1.3.11 implies that there exists a homotopy from $\mathsf{T}(1)^{-1}$ to I_n consisting of invertible matrices that do not depend on z. Therefore $[L] = [T][T(1)^{-1}] = [T]$, which establishes (i).

To show (ii), choose a natural number $M \geq m$ so large that $z^{M-m}L$ is a polynomial. Then $z^{M-m}L$ and $z^{M-m}I_n$ are in $\mathcal{P}^M(n, C(\mathcal{S}'X))$, and $[T] = [L] = [z^{M-m}L][z^{M-m}I_n]^{-1}$. □

Proposition 2.7.10 *Let X be a compact Hausdorff space. For each pair of natural numbers m and n with $m \geq 2$, there exists a continuous function*

$$\nu : \mathcal{P}^m(n, C(\mathcal{S}'X)) \longrightarrow \mathcal{P}^1((m+1)n, C(\mathcal{S}'X))$$

such that $[\nu(P)] = [P]$ in $K^{-1}(\mathcal{S}'X)$ for all P in $\mathcal{P}^m(n, C(\mathcal{S}'X))$.

Proof Write $P = \sum_{k=0}^m A_k z^k$, define

$$\tilde{\nu}(P) = \begin{pmatrix} A_0 & A_1 & A_2 & \cdots & A_{m-1} & A_m \\ -zI_n & I_n & 0 & \cdots & 0 & 0 \\ 0 & -zI_n & I_n & \cdots & 0 & 0 \\ \vdots & \vdots & \vdots & \ddots & \vdots & \vdots \\ 0 & 0 & 0 & \cdots & -zI_n & I_n \end{pmatrix},$$

and set $\nu(P)(z) = \tilde{\nu}(P)(z)(\tilde{\nu}(P)(1))^{-1}$. For $0 \leq t \leq 1$, define

$$R_t(z) = \begin{pmatrix} I_n & -t\sum_{k=1}^m A_k z^{k-1} & -t\sum_{k=2}^m A_k z^{k-2} & \cdots & -tA_m \\ 0 & I_n & 0 & \cdots & 0 \\ 0 & 0 & I_n & \cdots & 0 \\ \vdots & \vdots & \vdots & \ddots & \vdots \\ 0 & 0 & 0 & \cdots & I_n \end{pmatrix}.$$

This gives us a homotopy $\{(R_t(1))^{-1}R_t(z)\nu(P)\}$ in $\mathcal{P}^m((m+1)n, C(\mathcal{S}'X))$ from $\nu(P)$ to the matrix

$$\Gamma(z) = \begin{pmatrix} P & 0 & 0 & \cdots & 0 & 0 \\ -zI_n & I_n & 0 & \cdots & 0 & 0 \\ 0 & -zI_n & I_n & \cdots & 0 & 0 \\ \vdots & \vdots & \vdots & \ddots & \vdots & \vdots \\ 0 & 0 & 0 & \cdots & -zI_n & I_n \end{pmatrix}.$$

Define

$$H_t(z) = \begin{cases} R_{2t}(-1)R_{2t}(z)\nu(P)(z) & 0 \leq t \leq \frac{1}{2} \\ \Gamma((2-2t)z) & \frac{1}{2} \leq t \leq 1. \end{cases}$$

Then $\{H_t\}$ is a homotopy in $\mathcal{P}^m((m+1)n, C(S'X))$ from $\nu(\mathsf{P})$ to the matrix $\operatorname{diag}(\mathsf{P}, I_{mn})$, whence $[\nu(\mathsf{P})] = [\mathsf{P}]$ in $\mathrm{K}^{-1}(S'X)$.   □

Lemma 2.7.11 *Let X be a Hausdorff space, let n be a natural number, and suppose P is a polynomial in $\mathrm{GL}(n, C(X \times S^1))$. Then the Fourier series of P^{-1} converges uniformly to P^{-1}.*

Proof The inverse R of P is twice continuously differentiable as a function of θ (in fact, infinitely differentiable) because P has this property. Let B_k be the kth Fourier coefficient of R. For $k \neq 0$, integrate by parts twice to obtain

$$\|\mathsf{B}_k\|_\infty = \left\| \frac{-1}{2\pi k^2} \int_{-\pi}^{\pi} \mathsf{R}''(\theta) e^{ik\theta}\, d\theta \right\|_\infty \leq \frac{1}{k^2} \|\mathsf{R}''\|_\infty .$$

By the Weierstrauss M-test, the partial sums of the Fourier series for R form a Cauchy sequence and therefore converge uniformly to some function. We do not immediately know that the partial sums converge to R. However, Theorem 2.7.4 shows that the Cesàro means of R converge uniformly to R, and therefore the partial sums converge to R as well.   □

Proposition 2.7.10 suggests that we look more closely at polynomials of degree one.

Lemma 2.7.12 *Let X be a compact Hausdorff space, let n be a natural number, and suppose $\mathsf{P} = \mathsf{A}_0 + \mathsf{A}_1 z$ is an element of $\mathcal{P}^1(n, C(S'X))$. Let $\sum_{k \in \mathbb{Z}} \mathsf{B}_k z^k$ be the Fourier series of P^{-1}. Then the following hold:*

(i) $\mathsf{A}_0\mathsf{B}_k + \mathsf{A}_1\mathsf{B}_{k-1} = \mathsf{B}_k\mathsf{A}_0 + \mathsf{B}_{k-1}\mathsf{A}_1 = \begin{cases} I_n & \text{if } k = 0 \\ 0 & \text{if } k \neq 0; \end{cases}$

(ii) *Suppose that either $k < 0$ and $l \geq 0$, or that $l < 0$ and $k \geq 0$. Then:*

    (a) $\mathsf{B}_k\mathsf{A}_j\mathsf{B}_l = 0$ *for $j = 0, 1$*

    (b) $\mathsf{B}_k\mathsf{B}_l = 0$.

(iii) $\mathsf{B}_k\mathsf{B}_l = \mathsf{B}_l\mathsf{B}_k$ *for all integers k and l;*

(iv) $\mathsf{A}_j\mathsf{B}_k = \mathsf{B}_k\mathsf{A}_j$ *for $j = 0, 1$ and all integers k;*

(v) $\mathsf{A}_0\mathsf{A}_1 = \mathsf{A}_1\mathsf{A}_0;$

(vi) $\mathsf{A}_0\mathsf{B}_0$ *is an idempotent.*

Proof Multiply the Fourier series of P and P^{-1} together to obtain

$$I_n = PP^{-1} = (A_0 + A_1 z) \left(\sum_{k \in \mathbb{Z}} B_k z^k \right)$$

$$= \sum_{k \in \mathbb{Z}} (A_0 B_k + A_1 B_{k-1}) z^k,$$

from which the first part of (i) immediately follows. Similarly, the equation $P^{-1}P = I_n$ yields the second part of (i).

We verify (ii)(a) for the case when $j = 0$, $k < 0$, and $l \geq 0$; the other cases are similar. From (i), we have

$$B_k A_0 B_l = -B_{k-1} A_1 B_l = B_{k-1} A_0 B_{l+1}.$$

Iterating this process, we see that $B_k A_0 B_l = B_{k-r} A_0 B_{l+r}$ for all positive integers r. Because the Fourier series for P^{-1} converges absolutely, both B_{k-r} and B_{l+r} converge to 0 as r goes to ∞, and therefore $B_k A_0 B_l = 0$.

To show (ii)(b), note that $I_n = P(1) = A_0 + A_1$. Thus from (ii)(a) we obtain

$$B_k B_l = B_k (A_0 + A_1) B_l = B_k A_0 B_l + B_k A_1 B_l = 0 + 0 = 0.$$

In light of (ii)(b), we need only establish (iii) for the cases where $kl \geq 0$. Suppose that $l < k \leq 0$. We repeatedly use (i) to produce the equations

$$B_k A_0 B_l = -B_k A_1 B_{l-1} = B_{k+1} A_0 B_{l-1} = \cdots = B_l A_0 B_k$$
$$B_k A_1 B_l = -B_{k+1} A_0 B_l = B_{k+1} A_1 B_{l-1} = \cdots = B_l A_1 B_k.$$

Add these two equations to obtain

$$B_k B_l = B_k (A_0 + A_1) B_l = B_l (A_0 + A_1) B_k = B_l B_k.$$

The cases $k < l \leq 0$, $0 \leq k < l$, and $0 \leq l < k$ are similar.

To prove (iv), first take $k \leq 0$. From (i) and the fact that $A_0 + A_1 = I_n$, we have

$$A_0 B_k - B_k A_0 = -A_1 B_{k-1} + B_{k-1} A_1$$
$$= -(I_n - A_0) B_{k-1} + B_{k-1} (I_n - A_0)$$
$$= A_0 B_{k-1} - B_{k-1} A_0.$$

By iteration, we see that $A_0 B_k - B_k A_0 = A_0 B_{k-r} - B_{k-r} A_0$ for all positive integers r. Letting r go to infinity then shows that $A_0 B_k - B_k A_0 = 0$.

The case where $k > 0$ is similar. As for A_1, another application of (i) yields

$$A_1 B_k - B_k A_1 = -A_0 B_{k+1} + B_{k+1} A_0 = 0$$

for all integers k.

The fact that $A_1 = I_n - A_0$ immediately implies (v). Finally, to show that $A_0 B_0$ is an idempotent, observe that

$$A_0 B_0 A_0 B_0 = A_0 B_0 (I_n - A_1 B_{-1}) = A_0 B_0 - A_0 B_0 A_1 B_{-1} = A_0 B_0$$

by (i) and (ii)(a).   □

Proposition 2.7.13 *Let X be a compact Hausdorff space and let n be a natural number. There exists a continuous function*

$$\xi : \mathcal{P}^1(n, C(\mathcal{S}'X)) \longrightarrow \mathcal{I}(n, C(\mathcal{S}'X))$$

such that:

(i) *for every idempotent E in $\mathrm{M}(n, C(X))$,*

$$\xi(I_n - E + Ez) = I_n - E + Ez;$$

(ii) $[\xi(P)] = [P]$ *holds in $\mathrm{K}^{-1}(\mathcal{S}'X)$ for every P in $\mathcal{P}^1(n, C(\mathcal{S}'X))$.*

Proof Write $P = A_0 + A_1 z$ and $P^{-1} = \sum_{k \in \mathbb{Z}} B_k z^k$, and set $F = A_0 B_0$ and $E = I_n - F$. We define

$$\xi(P) = F + Ez.$$

The map ξ is continuous, because the Fourier coefficients of P and P^{-1} vary continuously. Part (i) of the proposition holds because the inverse of $I_n - E + Ez$ is $I_n - E + Ez^{-1}$.

Verifying (ii) is harder. Define $L_1(z) = E + P(z)F$ and $L_2(z) = F + P(z^{-1})Ez$. Parts (iv) and (v) of Lemma 2.7.12 imply that P commutes with F and E, and because $EF = FE = 0$, we obtain

$$\begin{aligned}
L_1(z) L_2(z^{-1})(\xi(P))(z) &= (E + P(z)F)\left(F + P(z)Ez^{-1}\right)(F + Ez) \\
&= \left(P(z)Ez^{-1} + P(z)F\right)(F + Ez) \\
&= P(z)(F + E) \\
&= P(z).
\end{aligned}$$

Furthermore, L_1 and L_2 are invertble, with $L_1^{-1}(z) = E + P^{-1}(z)F$ and $L_2^{-1}(z) = F + P^{-1}(z^{-1})Ez^{-1}$. Therefore, to prove the proposition, it suffices to show that there exists a homotopy $\{H_t\}$ in $GL(n, C(\mathcal{S}'X))$ from $L_1(z)L_2(z^{-1})$ to I_n.

We construct our homotopy by extending L_1 and L_2 and their inverses from functions on $X \times S^1$ to functions on the product of X and the closed unit disk. For $0 < t \le 1$ and z in S^1, we have

$$L_1^{-1}(tz) = E + P^{-1}(tz)F$$

$$= E + \sum_{k \in \mathbb{Z}} B_k A_0 B_0 (tz)^k$$

$$= E + B_0 F + \sum_{k=1}^{\infty} B_k A_0 B_0 (tz)^k;$$

the last equality follows from part (ii)(a) of Lemma 2.7.12. Because there are no negative powers of tz, our formula for $L_1^{-1}(tz)$ extends to $t = 0$, and $L_1(tz)L_1^{-1}(tz) = I_n$ for all $0 \le t \le 1$ and z in S^1. A similar computation yields

$$L_2^{-1}(tz^{-1}) = F + P^{-1}(t^{-1}z)Et^{-1}z$$

$$= F + \sum_{k \in \mathbb{Z}} B_k A_1 B_{-1} (t^{-1}z)^{k+1}$$

$$= F + B_{-1}E + \sum_{k=-\infty}^{-2} B_k A_1 B_{-1} (t^{-1}z)^{k+1}$$

$$= F + B_{-1}E + \sum_{j=1}^{\infty} B_{-(j+1)} A_1 B_{-1} (tz^{-1})^j.$$

Thus $L_2(tz^{-1})$ and $L_2^{-1}(tz^{-1})$ are defined for all $0 \le t \le 1$ and z in S^1, and $L_2(tz^{-1})L_2^{-1}(tz^{-1}) = I_n$. Define $H_t(z) = L_1(tz)L_2(tz^{-1})$. Then $\{H_t\}$ is the desired homotopy. □

We need one final lemma.

Lemma 2.7.14 *Let X be compact Hausdorff and let n be a natural number. Then the mapping $I_n - E + Ez \mapsto E$ is a continuous function from $\mathcal{I}(n, C(SX))$ to idempotents in $M(n, C(X))$.*

Proof The continuity of the map follows immediately from the equation

$$E = \frac{1}{2}\Big(I_n - \big(I_n - E + E(-1)\big)\Big).$$

□

Theorem 2.7.15 *For every compact Hausdorff space* X, *the map*

$$\beta : \mathrm{K}^0(X) \longrightarrow \mathrm{K}^{-1}(\mathcal{S}'X)$$

is a natural isomorphism of groups.

Proof We showed in Proposition 2.7.2 that β is a group homomorphism and a natural transformation, and the surjectivity of β follows immediately from Propositions 2.7.9, 2.7.10, and 2.7.13. To show that β is injective, suppose that

$$\beta([\mathsf{E}] - [\mathsf{F}]) = \beta[\mathsf{E}](\beta[\mathsf{F}])^{-1} = 1.$$

Choose a natural number n large enough so that E and F can be taken to be idempotents in $\mathrm{M}(n, C(X))$. By applying Lemma 2.7.7 and increasing n if necessary, we know there is a homotopy $\{\mathsf{J}_t\}$ in $\mathcal{L}^m(n, C(\mathcal{S}'X))$ from $I_n - \mathsf{E} + \mathsf{E}z$ to $I_n - \mathsf{F} + \mathsf{F}z$. Let m be the largest natural number that appears in any of the J_t, choose M so that $-M$ is the most negative power of z that appears in any of the J_t, and for $0 \leq t \leq 1$ define $\mathsf{P}_t = \mathrm{diag}(z^M \mathsf{J}_t, I_{Mn})$. Then $\{\mathsf{P}_t\}$ is a homotopy in $\mathcal{P}^{M+m}(n, C(\mathcal{S}'X))$ from $\mathrm{diag}(z^m(I_n - \mathsf{E} + \mathsf{E}z), I_{(M-1)n})$ to $\mathrm{diag}(z^m(I_n - \mathsf{F} + \mathsf{F}z), I_{(M-1)n})$. Repeated application of Proposition 1.3.3 gives us a homotopy $\{\mathsf{R}_t\}$ in $\mathcal{P}^{M+m}(n, C(\mathcal{S}'X))$ from $\mathrm{diag}(I_n - \mathsf{E} + \mathsf{E}z, I_{(M-1)n}z)$ to $\mathrm{diag}(z^m(I_n - \mathsf{E} + \mathsf{E}z), I_{(M-1)n})$.

Similarly, we can construct a homotopy $\{\mathsf{S}_t\}$ in $\mathcal{P}^{M+m}(n, C(\mathcal{S}'X))$ from $\mathrm{diag}(z^m(I_n - \mathsf{F} + \mathsf{F}z), I_{(M-1)n})$ to $\mathrm{diag}(I_n - \mathsf{F} + \mathsf{F}z, I_{(M-1)n})$. Define

$$\mathsf{H}_t = \begin{cases} \mathsf{R}_{1-3t} & 0 \leq t \leq \frac{1}{3} \\ \mathsf{P}_{3t-1} & \frac{1}{3} \leq t \leq \frac{2}{3} \\ \mathsf{S}_{3t-2} & \frac{2}{3} \leq t \leq 1. \end{cases}$$

Then $\{\xi\nu(\mathsf{H}_t)\}$ is a homotopy of idempotents from $\xi\nu(\mathrm{diag}(I_n - \mathsf{E} + \mathsf{E}z, I_{(M-1)n}z))$ to $\xi\nu(\mathrm{diag}(I_n - \mathsf{F} + \mathsf{F}z, I_{(M-1)n}z))$.

We can explicitly compute both ends of this homotopy. Note that

$$\mathrm{diag}(I_n - \mathsf{E} + \mathsf{E}z, I_{(M-1)n}z) = \mathrm{diag}(I_n - \mathsf{E}, 0_{(M-1)n}) + \mathrm{diag}(\mathsf{E}, I_{(M-1)n})z.$$

Let $\widetilde{\mathsf{A}}_0 = \mathrm{diag}(I_n - \mathsf{E}, 0_{(M-1)n})$ and $\widetilde{\mathsf{A}}_1 = \mathrm{diag}(\mathsf{E}, I_{(M-1)n})$. To simplify the notation, we drop the subscripts on identity matrices, and apply the formula for $\widetilde{\nu}$ in Proposition 2.7.10 to obtain

$$\widetilde{\nu}(\widetilde{A}_0 + \widetilde{A}_1 z) = \begin{pmatrix} \widetilde{A}_0 & \widetilde{A}_1 & 0 & 0 & 0 & \cdots & 0 & 0 \\ -Iz & I & 0 & 0 & 0 & \cdots & 0 & 0 \\ 0 & -Iz & I & 0 & 0 & \cdots & 0 & 0 \\ 0 & 0 & -Iz & I & 0 & \cdots & 0 & 0 \\ 0 & 0 & 0 & -Iz & I & \cdots & 0 & 0 \\ \vdots & \vdots & \vdots & \vdots & \vdots & \ddots & \vdots & \vdots \\ 0 & 0 & 0 & 0 & 0 & \cdots & -Iz & I \end{pmatrix}.$$

The equations $\widetilde{A}_0^2 = \widetilde{A}_0, \widetilde{A}_1^2 = \widetilde{A}_1, \widetilde{A}_0\widetilde{A}_1 = \widetilde{A}_1\widetilde{A}_0 = 0$ and $\widetilde{A}_0 + \widetilde{A}_1 = I$ imply

$$\left(\widetilde{\nu}(\widetilde{A}_0 + \widetilde{A}_1 z)(1)\right)^{-1} = \begin{pmatrix} I & -\widetilde{A}_1 & 0 & 0 & 0 & \cdots & 0 & 0 \\ I & \widetilde{A}_0 & 0 & 0 & 0 & \cdots & 0 & 0 \\ I & \widetilde{A}_0 & I & 0 & 0 & \cdots & 0 & 0 \\ I & \widetilde{A}_0 & I & I & 0 & \cdots & 0 & 0 \\ I & \widetilde{A}_0 & I & I & I & \cdots & 0 & 0 \\ \vdots & \vdots & \vdots & \vdots & \vdots & \ddots & \vdots & \vdots \\ I & \widetilde{A}_0 & I & I & I & \cdots & I & I \end{pmatrix}.$$

Let $\widetilde{z} = z - 1$. Then

$$\nu(\widetilde{A}_0 + \widetilde{A}_1 z) = \begin{pmatrix} I & 0 & 0 & 0 & 0 & \cdots & 0 & 0 \\ -I\widetilde{z} & \widetilde{A}_0 + \widetilde{A}_1 z & 0 & 0 & 0 & \cdots & 0 & 0 \\ -I\widetilde{z} & -\widetilde{A}_0\widetilde{z} & I & 0 & 0 & \cdots & 0 & 0 \\ -I\widetilde{z} & -\widetilde{A}_0\widetilde{z} & -I\widetilde{z} & I & 0 & \cdots & 0 & 0 \\ -I\widetilde{z} & -\widetilde{A}_0\widetilde{z} & -I\widetilde{z} & -I\widetilde{z} & I & \cdots & 0 & 0 \\ \vdots & \vdots & \vdots & \vdots & \vdots & \ddots & \vdots & \vdots \\ -I\widetilde{z} & -\widetilde{A}_0\widetilde{z} & -I\widetilde{z} & -I\widetilde{z} & -I\widetilde{z} & \cdots & -I\widetilde{z} & I \end{pmatrix},$$

which has inverse

$$\begin{pmatrix} I & 0 & 0 & \cdots & 0 & 0 \\ (\widetilde{A}_0 + \widetilde{A}_1 z^{-1})\widetilde{z} & \widetilde{A}_0 + \widetilde{A}_1 z^{-1} & 0 & \cdots & 0 & 0 \\ (\widetilde{A}_0 + \widetilde{A}_1 z^{-1})z\widetilde{z} & \widetilde{A}_0\widetilde{z} & I & \cdots & 0 & 0 \\ (\widetilde{A}_0 + \widetilde{A}_1 z^{-1})z^2\widetilde{z} & \widetilde{A}_0 z\widetilde{z} & I\widetilde{z} & \cdots & 0 & 0 \\ (\widetilde{A}_0 + \widetilde{A}_1 z^{-1})z^3\widetilde{z} & \widetilde{A}_0 z^2\widetilde{z} & Iz\widetilde{z} & \cdots & 0 & 0 \\ \vdots & \vdots & \vdots & \ddots & \vdots & \vdots \\ (\widetilde{A}_0 + \widetilde{A}_1 z^{-1})z^{m-1}\widetilde{z} & \widetilde{A}_0 z^{m-2}\widetilde{z} & Iz^{m-3}\widetilde{z} & \cdots & I\widetilde{z} & I \end{pmatrix}.$$

In the notation of Lemma 2.7.12, we have

$$
A_0 = \begin{pmatrix}
I & 0 & 0 & 0 & 0 & \cdots & 0 & 0 \\
I & \tilde{A}_0 & 0 & 0 & 0 & \cdots & 0 & 0 \\
I & \tilde{A}_0 & I & 0 & 0 & \cdots & 0 & 0 \\
I & \tilde{A}_0 & I & I & 0 & \cdots & 0 & 0 \\
I & \tilde{A}_0 & I & I & I & \cdots & 0 & 0 \\
\vdots & \vdots & \vdots & \vdots & \vdots & \ddots & \vdots & \vdots \\
I & \tilde{A}_0 & I & I & I & \cdots & I & I
\end{pmatrix}
$$

and

$$
B_0 = \begin{pmatrix}
I & 0 & 0 & 0 & 0 & \cdots & 0 & 0 \\
\tilde{A}_1 - \tilde{A}_0 & \tilde{A}_0 & 0 & 0 & 0 & \cdots & 0 & 0 \\
-\tilde{A}_1 & -\tilde{A}_0 & I & 0 & 0 & \cdots & 0 & 0 \\
0 & 0 & -I & I & 0 & \cdots & 0 & 0 \\
0 & 0 & 0 & -I & I & \cdots & 0 & 0 \\
\vdots & \vdots & \vdots & \vdots & \vdots & \ddots & \vdots & \vdots \\
0 & 0 & 0 & 0 & 0 & \cdots & -I & I
\end{pmatrix}.
$$

Thus

$$
A_0 B_0 = \mathrm{diag}\left(\begin{pmatrix} I_{Mn} & 0 \\ \tilde{A}_1 & \tilde{A}_0 \end{pmatrix}, I_{Mn(m-2)} \right)
$$

$$
= \mathrm{diag}\left(\begin{pmatrix} I_{Mn} & 0 \\ \mathrm{diag}(E, I_{(M-1)n}) & \mathrm{diag}(I_n - E, 0_{(M-1)n}) \end{pmatrix}, I_{Mn(m-2)} \right).
$$

A similar computation with F gives us

$$
\mathrm{diag}\left(\begin{pmatrix} I_{Mn} & 0 \\ \mathrm{diag}(F, I_{(M-1)n}) & \mathrm{diag}(I_n - F, 0_{(M-1)n}) \end{pmatrix}, I_{Mn(m-2)} \right)
$$

at the other end of our homotopy of idempotents. Lemma 2.7.14 then implies that

$$
\left[\mathrm{diag}\left(\begin{pmatrix} I_{Mn} & 0 \\ \mathrm{diag}(E, I_{(M-1)n}) & \mathrm{diag}(I_n - E, 0_{(M-1)n}) \end{pmatrix}, I_{Mn(m-2)} \right) \right] =
$$

$$
\left[\mathrm{diag}\left(\begin{pmatrix} I_{Mn} & 0 \\ \mathrm{diag}(F, I_{(M-1)n}) & \mathrm{diag}(I_n - F, 0_{(M-1)n}) \end{pmatrix}, I_{Mn(m-2)} \right) \right]
$$

in $K^0(X)$, whence

$$\left[\begin{pmatrix} I_{Mn} & 0 \\ \mathrm{diag}(\mathsf{E}, I_{(M-1)n}) & \mathrm{diag}(I_n - \mathsf{E}, 0_{(M-1)n}) \end{pmatrix}\right] = \\ \left[\begin{pmatrix} I_{Mn} & 0 \\ \mathrm{diag}(\mathsf{F}, I_{(M-1)n}) & \mathrm{diag}(I_n - \mathsf{F}, 0_{(M-1)n}) \end{pmatrix}\right].$$

Conjugate the left side of this equation by

$$\begin{pmatrix} I_{Mn} & \mathrm{diag}(I_n - \mathsf{E}, 0_{(M-1)n}) \\ -\mathrm{diag}(\mathsf{E}, I_{(M-1)n}) & I_{Mn} \end{pmatrix}$$

and the right side by

$$\begin{pmatrix} I_{Mn} & \mathrm{diag}(I_n - \mathsf{F}, 0_{(M-1)n}) \\ -\mathrm{diag}(\mathsf{F}, I_{(M-1)n}) & I_{Mn} \end{pmatrix}$$

to obtain

$$\left[\mathrm{diag}\left(I_{Mn}, I_n - \mathsf{E}, 0_{(M-1)n}\right)\right] = \left[\mathrm{diag}\left(I_{Mn}, I_n - \mathsf{F}, 0_{(M-1)n}\right)\right].$$

Finally, using Lemma 1.7.3 and the definition of addition in $K^0(X)$, we obtain

$$[I_{Mn}] + [I_n - \mathsf{E}] + [0_{(M-1)n}] = [I_{Mn}] + [I_n - \mathsf{F}] + [0_{(M-1)n}]$$
$$[I_n - \mathsf{E}] = [I_n - \mathsf{F}]$$
$$[I_n] - [\mathsf{E}] = [I_n] - [\mathsf{F}]$$
$$[\mathsf{E}] = [\mathsf{F}].$$

$\square$

Corollary 2.7.16 *For every locally compact Hausdorff space X, there exists a natural isomorphism $\beta : K^0(X) \longrightarrow K^{-2}(X)$.*

Proof Consider the commutative diagram

$$\begin{array}{ccccccccc} 0 & \longrightarrow & K^0(X) & \longrightarrow & K^0(X^+) & \longrightarrow & K^0(\{\infty\}) & \longrightarrow & 0 \\ & & & & \downarrow{\scriptstyle\beta} & & \downarrow{\scriptstyle\beta} & & \\ 0 & \longrightarrow & K^{-2}(X) & \longrightarrow & K^{-2}(X^+) & \longrightarrow & K^{-2}(\{\infty\}) & \longrightarrow & 0 \end{array}$$

that has exact rows. The inclusion of $\{\infty\}$ into X^+ makes the top row split exact by Theorem 2.6.11. Moreover, if we take each topological space in the top row and form its Cartesian product with $\mathbb{R}^2$, Definition

2.6.14 and Theorem 2.6.11 yield that the bottom row is also split exact. The desired result is then a consequence of Proposition 1.8.9. □

Corollary 2.7.17 *Suppose that A is a closed subspace of a locally compact Hausdorff space X, and let j denote the inclusion of A into X. Then there exists an exact sequence*

$$
\begin{array}{ccccc}
K^0(X \backslash A) & \longrightarrow & K^0(X) & \xrightarrow{\ j^* \ } & K^0(A) \\
{\scriptstyle \partial} \uparrow & & & & \downarrow {\scriptstyle \partial} \\
K^{-1}(A) & \xleftarrow{\ j^* \ } & K^{-1}(X) & \longleftarrow & K^{-1}(X \backslash A),
\end{array}
$$

and both connecting maps ∂ are natural.

Proof We know from Theorem 2.6.16 that the map $\partial : K^{-1}(A) \longrightarrow K^0(X, A)$ is natural and that the sequence is exact at the groups $K^{-1}(X)$, $K^{-1}(A)$, $K^0(X \backslash A)$, and $K^0(X)$. Exactness at $K^0(A)$ and $K^{-1}(X \backslash A)$ and the naturality of the other connecting map are immediate consequences of Theorems 2.6.16 and 2.7.15. □

Proposition 2.7.18 *Let A be a closed subspace of a locally compact Hausdorff space X and let $\partial : K^0(A) \longrightarrow K^{-1}(X \backslash A)$ be the connecting homomorphism in Corollary 2.7.17. Suppose that E is an idempotent in $M(n, C(A^+))$ such that $\mathsf{E}(\infty) = \mathrm{diag}(I_k, 0_{n-k})$. If B is an element (not necessarily an idempotent) in $M(n, C(X^+))$ whose restriction to A^+ is E, then*

$$\partial\big([\mathsf{E}] - [I_k]\big) = [\exp(-2\pi i \mathsf{B})]$$

in $K^{-1}(X \backslash A)$.

Proof We express $\partial : K^0(A) \longrightarrow K^{-1}(X \backslash A)$ as a composition of maps. The first comes from identifying $(0, 1)$ with $S^1 \backslash \{1\}$ via the homeomorphism $t \mapsto e^{2\pi i t}$ and applying the Bott periodicity isomorphism

$$\beta : K^0(A) \longrightarrow K^{-1}(A \times (0, 1)).$$

The second homomorphism is the connecting map

$$\partial' : K^{-1}(A \times (0, 1)) \longrightarrow K^0\big((X \backslash A) \times (0, 1)\big)$$

that arises when we apply Theorem 2.6.12 to the inclusion of $A \times (0, 1)$

into $X \times (0,1)$. The third map is the isomorphism

$$\sigma : K^{-1}(X \backslash A) \longrightarrow K^0\big((X \backslash A) \times (0,1)\big)$$

that comes from Theorem 2.6.13 and the identification on $\mathbb{R}$ with $(0,1)$. Then $\partial = \sigma^{-1}\partial'\beta$; in other words, $\partial : K^0(A) \longrightarrow K^{-1}(X \backslash A)$ is the unique homomorphism that makes the square

$$
\begin{array}{ccc}
K^0(A) & \xrightarrow{\ \partial\ } & K^{-1}(X \backslash A) \\
\Big\downarrow{\scriptstyle\beta} & & \Big\downarrow{\scriptstyle\sigma} \\
K^{-1}(A \times (0,1)) & \xrightarrow{\ \partial'\ } & K^0\big((X \backslash A) \times (0,1)\big)
\end{array}
$$

commute.

We have

$$
\begin{aligned}
\beta\big([\mathsf{E}] - [I_k]\big) &= [I_n - \mathsf{E} + e^{2\pi i t}\mathsf{E}]\,[(I_k - I_k + e^{2\pi i t}I_k)^{-1}] \\
&= [I_n - \mathsf{E} + e^{2\pi i t}\mathsf{E}]\,[e^{-2\pi i t}I_k].
\end{aligned}
$$

For all $0 \le t \le 1$, we obtain

$$
\begin{aligned}
\exp(2\pi i t\mathsf{E}) &= I_n + \sum_{k=1}^{\infty} \frac{(2\pi i t)^k}{k!}\mathsf{E}^k \\
&= I_n + \sum_{k=1}^{\infty} \frac{(2\pi i t)^k}{k!}\mathsf{E} \\
&= I_n - \mathsf{E} + \sum_{k=0}^{\infty} \frac{(2\pi i t)^k}{k!}\mathsf{E}^k \\
&= I_n - \mathsf{E} + e^{2\pi i t}\mathsf{E},
\end{aligned}
$$

and thus

$$
\begin{aligned}
\beta\big([\mathsf{E}] - [I_k]\big) &= [\exp(2\pi i t\mathsf{E})]\,[e^{-2\pi i t}I_k] \\
&= [\exp(2\pi i t\mathsf{E})]\,[\mathrm{diag}(e^{-2\pi i t}I_k, I_{n-k})] \\
&= [\exp(2\pi i t\mathsf{E})\,\mathrm{diag}(e^{-2\pi i t}I_k, I_{n-k})].
\end{aligned}
$$

Note that

$$\exp(2\pi i t\mathsf{E})\,\mathrm{diag}(e^{-2\pi i t}I_k, I_{n-k}) = I_n$$

when t equals 0 or 1, and at ∞ we have

$$
\begin{aligned}
\exp(2\pi it\mathsf{E})\,\mathrm{diag}(e^{-2\pi it}I_k, I_{n-k}) \\
&= \exp(2\pi it\,\mathrm{diag}(I_k, 0_{n-k}))\,\mathrm{diag}(e^{-2\pi it}I_k, I_{n-k}) \\
&= \mathrm{diag}(e^{2\pi it}I_k, I_{n-k})\,\mathrm{diag}(e^{-2\pi it}I_k, I_{n-k}) \\
&= I_n.
\end{aligned}
$$

Therefore $\exp(2\pi it\mathsf{E})\,\mathrm{diag}(e^{-2\pi it}I_k, I_{n-k})$ determines an element of $\mathrm{GL}(n, C(A^+ \times (0,1)^+))$, and hence also an invertible matrix over the one-point compactification of $A \times (0,1)$, because $(A \times (0,1))^+$ is homeomorphic to $A^+ \times (0,1)^+$ with $(\{\infty_A\} \times (0,1)) \cup (A \times \{\infty_{(0,1)}\})$ collapsed to a point. Use Corollary 1.3.17 to produce T in $\mathrm{GL}(2n, C((X \times (0,1))^+))$ with the property that

$$
\mathsf{T}|(A \times (0,1))^+ =
$$
$$
\mathrm{diag}\big(\exp(2\pi it\mathsf{E})\,\mathrm{diag}(e^{-2\pi it}I_k, I_{n-k}), \exp(-2\pi it\mathsf{E})\,\mathrm{diag}(e^{2\pi it}I_k, I_{n-k})\big).
$$

Then $\partial'\beta'([\mathsf{E}] - [I_k]) = [\mathsf{T}\,\mathrm{diag}(I_n, 0_n)\mathsf{T}^{-1}] - [I_n]$.

The next step is to choose an element B in $\mathrm{M}(n, C(X+))$ whose restriction to $\mathrm{M}(n, C(A^+))$ is E; we can find such an element by applying the Tietze extension theorem to each entry of E. When restricted to A, the matrix $\exp(-2\pi i\mathsf{B})$ equals I_n, and therefore makes sense as an element of $\mathrm{GL}(n, C((X \backslash A)^+))$. View the matrix T that we constructed in the previous paragraph as a matrix over $X \times [0,1]$ that is constant on the set $X \times \{0,1\}$, and define

$$
\widetilde{\mathsf{T}} = \mathsf{T}\,\mathrm{diag}(\exp(-2\pi it\mathsf{B}), \exp(2\pi it\mathsf{B})).
$$

When $t = 0$, the matrix T has the constant value I_{2n} because it has that value when restricted to $A \times \{0\}$, and thus $\widetilde{\mathsf{T}}|(X \times \{0\}) = I_{2n}$. Moreover, a simple computation shows that $\widetilde{\mathsf{T}}|(A \times (0,1]) = I_{2n}$ as well. Therefore Theorem 2.6.13 gives us

$$
\begin{aligned}
\sigma[\exp(-2\pi i\mathsf{B})] &= [\widetilde{\mathsf{T}}\,\mathrm{diag}(I_n, 0_n)\widetilde{\mathsf{T}}^{-1}] - [I_n] \\
&= [\mathsf{T}\,\mathrm{diag}(I_n, 0_n)\mathsf{T}^{-1}] - [I_n] \\
&= \partial'\beta([\mathsf{E}] - [I_k]),
\end{aligned}
$$

as desired. $\qquad\square$

Corollary 2.7.19 *For every compact pair* (X, A), *there exists an exact* *sequence*

$$K^0(X, A) \xrightarrow{\mu_1^*} K^0(X) \xrightarrow{j*} K^0(A)$$

$$\partial \uparrow \qquad\qquad\qquad\qquad \downarrow \partial$$

$$K^{-1}(A) \xleftarrow{j^*} K^{-1}(X) \xleftarrow{\mu_1^*} K^{-1}(X, A)$$

whose connecting homomorphisms ∂ are natural.

Proof The naturality of $\partial : K^{-1}(A) \longrightarrow K^0(X, A)$ and the exactness of the sequence at $K^{-1}(A)$, $K^{-1}(X)$, $K^0(X, A)$, and $K^0(X)$ come from Theorem 2.4.1, and remainder of the result follows from Theorem 2.3.17, Theorem 2.6.11, and Corollary 2.7.17. Theorem 2.3.17 and Proposition 2.7.18 imply that in the notation of Proposition 2.7.18, the connecting homomorphism $\partial : K^0(A) \longrightarrow K^{-1}(X, A)$ is given by the formula

$$\partial([\mathsf{E}] - [I_k]) = [\exp(-2\pi i \mathsf{B}), I_n].$$

$\square$

We close this section by noting that we can now define K^{p} for every integer p.

Definition 2.7.20 *Let p be any integer.*

(i) *For every locally compact Hausdorff space X, define*

$$K^{\mathsf{p}}(X) = \begin{cases} K^0(X) & \text{if } p \text{ is even} \\ K^{-1}(X) & \text{if } p \text{ is odd.} \end{cases}$$

(ii) *For every compact pair (X, A), define*

$$K^{\mathsf{p}}(X, A) = \begin{cases} K^0(X \backslash A) & \text{if } p \text{ is even} \\ K^{-1}(X \backslash A) & \text{if } p \text{ is odd.} \end{cases}$$

2.8 Computation of some K groups

We are finally ready to compute the K-theory groups of some interesting topological spaces.

Example 2.8.1 (*K*-theory of $\mathbb{R}^n$ **and** S^n) *For each natural number* n*, Theorem 2.7.15 gives us*

$$\mathrm{K}^0(\mathbb{R}^n) \cong \mathrm{K}^0((pt) \times \mathbb{R}^n) = \mathrm{K}^{-\mathrm{n}}(pt) \cong \begin{cases} \mathbb{Z} & \text{if } n \text{ is even} \\ 0 & \text{if } n \text{ is odd.} \end{cases}$$

This in turn yields

$$\mathrm{K}^{-1}(\mathbb{R}^n) \cong \mathrm{K}^0(\mathbb{R}^{n+1}) \cong \begin{cases} 0 & \text{if } n \text{ is even} \\ \mathbb{Z} & \text{if } n \text{ is odd.} \end{cases}$$

Combining these isomorphisms with Definition 2.7.20, we obtain

$$\mathrm{K}^{\mathrm{p}}(\mathbb{R}^n) \cong \begin{cases} \mathbb{Z} & \text{if } p + n \text{ is even} \\ 0 & \text{if } p + n \text{ is odd} \end{cases}$$

for all natural numbers n *and integers* p*.*

  The n*-sphere* S^n *is the one-point compactification of* $\mathbb{R}^n$*. Thus*

$$\widetilde{\mathrm{K}}^0(S^n) = \widetilde{\mathrm{K}}^0\big((\mathbb{R}^n)^+\big) = \mathrm{K}^0(\mathbb{R}^n),$$

and therefore

$$\mathrm{K}^0(S^n) \cong \begin{cases} \mathbb{Z} \oplus \mathbb{Z} & \text{if } n \text{ is even} \\ \mathbb{Z} & \text{if } n \text{ is odd.} \end{cases}$$

Similarly, $\widetilde{\mathrm{K}}^{-1}(S^n) \cong \mathrm{K}^{-1}(\mathbb{R}^n)$*, and therefore*

$$\mathrm{K}^{-1}(S^n) \cong \begin{cases} 0 & \text{if } n \text{ is even} \\ \mathbb{Z} & \text{if } n \text{ is odd.} \end{cases}$$

  Because the group $\mathrm{K}^0(S^1)$ *is isomorphic to* $\mathbb{Z}$*, the trivial line bundle* $\Theta^1(S^1)$ *is a generator. On the other hand, the topological space* $\mathcal{S}'(pt)$ *is homeomorphic to* S^1*, so the group* $\mathrm{K}^{-1}(S^1)$ *is generated by*

$$\beta[\Theta^1(S^1)] = \beta[1] = [(1-1) + 1z] = [z].$$

If we represent $\mathbb{R}$ *as* $S^1 \backslash \{1\}$*, then* $[z]$ *serves as a generator of* $\mathrm{K}^{-1}(\mathbb{R})$ *as well.*

  We can also write down generators for the groups $\mathrm{K}^0(S^2)$ *and* $\mathrm{K}^0(\mathbb{R}^2)$*. We have the isomorphism* $\mathrm{K}^0(S^2) \cong \mathrm{K}^0(\mathbb{R}^2) \oplus \mathrm{K}^0(pt)$*; the trivial bundles generate the second summand. To find a generator of* $\mathrm{K}^0(\mathbb{R}^2)$*, apply*

Theorem 2.7.19 to the compact pair (B^2, S^1) to obtain the exact sequence

$$K^0(B^2, S^1) \xrightarrow{\mu_1^*} K^0(B^2) \xrightarrow{j*} K^0(S^1)$$

$$\Big\uparrow \partial \qquad\qquad\qquad\qquad\qquad \Big\downarrow \partial$$

$$K^{-1}(S^1) \xleftarrow{\quad j* \quad} K^{-1}(B^2) \xleftarrow{\mu_1^*} K^{-1}(B^2, S^1).$$

The disk B^2 is contractible, so $K^0(B^2) \cong \mathbb{Z}$ and the map j^* from $K^0(B^2)$ to $K^0(S^1)$ is an isomorphism. Thus $\mu_1^* : K^0(B^2, S^1) \longrightarrow K^0(B^2)$ is zero. The contractibility of B^2 yields that $K^{-1}(B^2)$ is trivial and therefore the connecting map $\partial : K^{-1}(S^1) \longrightarrow K^0(B^2, S^1)$ is an isomorphism.

For each integer k, define

$$\mathsf{T} = \begin{pmatrix} z^k & -\sqrt{1 - |z|^{2k}} \\ \sqrt{1 - |z|^{2k}} & \bar{z}^k \end{pmatrix}.$$

The matrix T is in $\mathrm{GL}(2, C(B^2))$ because $\det \mathsf{T} = |z|^{2k} + (1 - |z|^{2k}) = 1$ for all z in B^2. Furthermore, the restriction of T to S^1 is $\mathrm{diag}(z^k, \bar{z}^k) = \mathrm{diag}(z^k, z^{-k})$. Theorem 2.4.1 gives us

$$\partial[z^k] = \left[\mathsf{T}\,\mathrm{diag}(1,0)\mathsf{T}^{-1}, \mathrm{diag}(1,0) \right] - [1,1]$$

$$= \left[\begin{pmatrix} |z|^{2k} & z^k\sqrt{1 - |z|^{2k}} \\ \bar{z}^k\sqrt{1 - |z|^{2k}} & 1 - |z|^{2k} \end{pmatrix}, \begin{pmatrix} 1 & 0 \\ 0 & 0 \end{pmatrix} \right] - [1,1],$$

and thus every element of $K^0(B^2 \backslash S^1)$ can be uniquely written in the form

$$\left[\begin{pmatrix} |z|^{2k} & z^k\sqrt{1 - |z|^{2k}} \\ \bar{z}^k\sqrt{1 - |z|^{2k}} & 1 - |z|^{2k} \end{pmatrix} \right] - [1]$$

for some integer k. Now, the function $\phi(z) = z(1 + |z|^2)^{-1/2}$ is a homeomorphism from $\mathbb{C} \cong \mathbb{R}^2$ to $B^2 \backslash S^1$, and if we take $k = 1$ we obtain a generator

$$\mathbf{b} = \phi^* \left(\left[\begin{pmatrix} |z|^2 & z\sqrt{1 - |z|^2} \\ \bar{z}\sqrt{1 - |z|^2} & 1 - |z|^2 \end{pmatrix} \right] - [1] \right)$$

$$= \frac{1}{|z|^2 + 1} \left[\begin{pmatrix} |z|^2 & z \\ \bar{z} & 1 \end{pmatrix} \right] - [1]$$

of $K^0(\mathbb{R}^2)$.

Example 2.8.2 (*K*-theory of $\mathbb{CP}^1$) *Because* $\mathbb{CP}^1$ *is homeomorphic to* S^2, *the previous example computes the K-theory of* $\mathbb{CP}^1$ *up to isomorphism. To explicitly write down a nontrivial generator of* $\mathrm{K}^0(\mathbb{CP}^n)$, *identify* S^2 *with the one-point compactification of* $\mathbb{C}$. *Then the map* $\psi : \mathbb{CP}^1 \longrightarrow S^2$ *given by* $\psi([z_1, z_2]) = z_1/z_2$ *is a homeomorphism and*

$$\psi^*(\mathbf{b}) = \left[\frac{1}{|z_1|^2 + |z_2|^2} \begin{pmatrix} |z_1|^2 & z_1 \bar{z}_2 \\ \bar{z}_1 z_2 & |z_2|^2 \end{pmatrix} \right] - [1]$$

in the notation of Example 2.8.1. Via the monoid isomorphism between $\mathrm{Idem}(\mathbb{CP}^1)$ *and* $\mathrm{Vect}(\mathbb{CP}^1)$, *we can express* $\psi^*(\mathbf{b})$ *as* $[H^*] - [1]$, *where* $[H^*]$ *is the vector bundle we constructed in Example 1.5.10.*

Definition 2.8.3 *The* wedge *of pointed spaces* (X, x_0) *and* (Y, y_0) *is the pointed space*

$$X \vee Y = (X \times \{y_0\}) \cup (\{x_0\} \times Y) \subseteq X \times Y$$

endowed with the subspace topology and with (x_0, y_0) *as basepoint.*

Example 2.8.4 (*K*-theory of the figure eight) *In this example we compute the K-theory of the "figure eight" space* $S^1 \vee S^1$. *Let* j *be the inclusion of one of the copies of* S^1 *into* $S^1 \vee S^1$ *and define a continuous function* $\psi : S^1 \vee S^1 \longrightarrow S^1$ *by identifying the two copies of* S^1 *to a single circle. Then* ψj *is the identity map on* S^1. *The topological space* $(S^1 \vee S^1) \backslash S^1$ *is homeomorphic to* $\mathbb{R}$, *and hence Theorem 2.6.11 and Example 2.8.1 yield*

$$\mathrm{K}^0(S^1 \vee S^1) \cong \mathrm{K}^0(S^1) \oplus \mathrm{K}^0(\mathbb{R}) \cong \mathbb{Z}$$
$$\mathrm{K}^{-1}(S^1 \vee S^1) \cong \mathrm{K}^{-1}(S^1) \oplus \mathrm{K}^{-1}(\mathbb{R}) \cong \mathbb{Z} \oplus \mathbb{Z}.$$

The group $\mathrm{K}^0(S^1 \vee S^1)$ *is generated by trivial bundles. For each pair of integers* (k, l), *define* $f_{(k,l)}$ *on* $S^1 \vee S^1$ *to be the function that is* z^k *on one circle and* z^l *on the other circle. Then the map* $(k, l) \mapsto [f_{(k,l)}]$ *is an isomorphism from* $\mathbb{Z} \oplus \mathbb{Z}$ *to* $\mathrm{K}^{-1}(S^1 \vee S^1)$.

Example 2.8.5 (*K*-theory of the torus) *Let* j *be the inclusion map from* $S^1 \times \{1\} \cong S^1$ *into the torus* $S^1 \times S^1$ *and let* ψ *be the projection of* $S^1 \times S^1$ *onto* $S^1 \times \{1\}$. *Then* ψj *is the identity. The complement of* $S^1 \times \{1\}$ *in* $S^1 \times S^1$ *is homeomorphic to* $S^1 \times \mathbb{R}$, *whence*

$$\mathrm{K}^0(S^1 \times S^1) \cong \mathrm{K}^0(S^1) \oplus \mathrm{K}^0(S^1 \times \mathbb{R}) \cong \mathrm{K}^0(S^1) \oplus \mathrm{K}^{-1}(S^1) \cong \mathbb{Z} \oplus \mathbb{Z}$$
$$\mathrm{K}^{-1}(S^1 \times S^1) \cong \mathrm{K}^{-1}(S^1) \oplus \mathrm{K}^{-1}(S^1 \times \mathbb{R}) \cong \mathrm{K}^{-1}(S^1) \oplus \mathrm{K}^{-2}(S^1) \cong \mathbb{Z} \oplus \mathbb{Z}.$$

Example 2.8.6 (*K-theory of the real projective plane*) *The* real projective plane *is the compact Hausdorff quotient space* $\mathbb{RP}^2$ *constructed by taking the closed unit disk* B^2 *and identifying antipodal points on* S^1; *in other words, we identify* z *with* $-z$ *when* $|z| = 1$. *Let* $q : B^2 \longrightarrow \mathbb{RP}^2$ *be the quotient map and let* $A = q(S^1)$; *we represent matrices over* A *as matrices over* S^1 *that are equal at antipodal points. Corollary 2.7.17 gives us the exact sequence*

$$
\begin{array}{ccc}
K^0(\mathbb{RP}^2\backslash A) \longrightarrow K^0(\mathbb{RP}^2) \xrightarrow{\ j^*\ } K^0(A) \\
\big\uparrow\partial \hspace{6cm} \big\downarrow\partial \\
K^{\text{-}1}(A) \xleftarrow{\ j^*\ } K^{\text{-}1}(\mathbb{RP}^2) \longleftarrow K^{\text{-}1}(\mathbb{RP}^2\backslash A).
\end{array}
$$

The map $\phi : A \longrightarrow S^1$ *given by* $\phi(z) = z^2$ *is well defined and is a homeomorphism, so* $K^0(A)$ *and* $K^{\text{-}1}(A)$ *are both isomorphic to* $\mathbb{Z}$. *The complement of* A *in* $\mathbb{RP}^2$ *is a homeomorphic to* $\mathbb{R}^2$, *so we also know that* $K^0(\mathbb{RP}^2\backslash A)$ *is isomorphic to* $\mathbb{Z}$ *and that* $K^{\text{-}1}(\mathbb{RP}^2\backslash A)$ *is trivial.*

To compute the K-theory of $\mathbb{RP}^2$, *we examine the connecting homomorphism*

$$\partial : K^{\text{-}1}(A) \longrightarrow K^0(\mathbb{RP}^2\backslash A).$$

Example 2.8.1 tells us that $[z]$ *is a generator* $K^{\text{-}1}(S^1)$, *so* $q[z] = [z^2]$ *is a generator of* $K^{\text{-}1}(A)$. *The matrix* $\mathrm{diag}(z^2, z^{-2})$ *over* A *lifts to the invertible matrix*

$$\mathsf{T} = \begin{pmatrix} z^2 & -\sqrt{1 - |z|^4} \\ \sqrt{1 - |z|^4} & \bar{z}^2 \end{pmatrix}$$

over $\mathbb{RP}^2$, *and Theorem 2.4.1 yields*

$$\partial[z^2] = \left[\begin{pmatrix} |z|^4 & z^2\sqrt{1 - |z|^4} \\ \bar{z}^2\sqrt{1 - |z|^4} & 1 - |z|^4 \end{pmatrix} \right] - [1].$$

From Example 2.8.1, we see that $\partial[z]$ *is two times a generator of the group* $K^0(\mathbb{RP}^2\backslash A)$. *Therefore* $K^0(\mathbb{RP}^2) \cong \mathbb{Z}\oplus\mathbb{Z}_2$ *and* $K^{\text{-}1}(\mathbb{RP}^2)$ *is trivial.*

Example 2.8.6 shows that it is possible for K-theory groups to have torsion.

2.9 Cohomology theories and K-theory

K-theory is an example of a *cohomology theory*:

Definition 2.9.1 *A cohomology theory with coefficients in $\mathbb{Z}$ is a doubly infinite sequence $H^* = \{H^p\}_{p \in \mathbb{Z}}$ of contravariant functors from the category of compact pairs to the category of abelian groups such that the Eilenberg–Steenrod axioms hold:*

- Homotopy axiom: *If $\{f_t\}$ is a homotopy of morphisms from one compact pair to another, then $H^p(f_0) = H^p(f_1)$ for all p.*
- Exactness axiom: *For any compact pair (X, A), the inclusion of A into X induces a long exact sequence*

$$\cdots \xrightarrow{\partial} H^p(X, A) \longrightarrow H^p(X) \longrightarrow$$

$$H^p(A) \xrightarrow{\partial} H^{p+1}(X, A) \longrightarrow \cdots$$

with natural connecting maps.
- Excision axiom: *Given a compact pair (X, A) and open subset U of X whose closure is in the interior of A, the inclusion morphism from $(X \backslash U, A \backslash U)$ to (X, A) induces an isomorphism from $H^p(X, A)$ to $H^p(X \backslash U, A \backslash U)$ for all p.*

If a cohomology theory satisfies

- Dimension axiom:

$$H^p(pt) \cong \begin{cases} \mathbb{Z} & \text{if } p = 0 \\ 0 & \text{if } p \neq 0 \end{cases},$$

we say that H^* is an ordinary *cohomology theory; otherwise H^* is called an* extraordinary *cohomology theory.*

We have shown in this chapter that K-theory is an extraordinary cohomology theory. All ordinary cohomology theories agree, up to isomorphism, on a large subcategory of the category of compact pairs; this subcategory covers most topological spaces of interest. In contrast, K-theory is only one of many extraordinary cohomology theories. We will explore the connection between K-theory and ordinary cohomology theories in Chapter 4.

2.10 Notes

The original definition of K-theory for topological spaces goes back to Atiyah and Hirzebruch [4]. Our definition for K-theory of compact pairs largely comes from [15]; the reader can find alternate definitions in [12] and [3]. There are many proofs of Bott periodicity in the literature; ours contains elements of the proofs in [12] and [16], which are in turn are variations on Atiyah's original proof in [3]. For more information about cohomology theories and the Eilenberg–Steenrod axioms, the reader should consult [8].

Exercises

2.1   Compute the K-theory groups of

   (a) $\bigoplus$
   (b) the wedge of n copies of S^1
   (c) $S^1 \vee S^1 \vee S^2$
   (d) $\mathbb{R}^n \backslash \{0\}$
   (e) the closed Möbius band; i.e., the quotient space obtained by taking $[0,1] \times [0,1]$ and identifying $(1,t)$ with $(0, 1-t)$ for all $0 \le t \le 1$
   (f) the open Möbius band; i.e., the closed Möbius band with its boundary removed
   (g) the Klein bottle

2.2   Write the 3-sphere S^3 as the subspace

$$S^3 = \{(z,w) \in \mathbb{C}^2 : |z|^2 + |w|^2 = 1\}$$

   of $\mathbb{C}^2$. For each natural number n, let $L(n)$ denote the quotient space constructed by identifying (z,w) and $(e^{2\pi i/n}z, e^{2\pi ki/n}w)$ for each point (z,w) in S^3. Show that $\mathrm{K}^0(L(n)) \cong \mathbb{Z} \oplus \mathbb{Z}_p$ and that $\mathrm{K}^{-1}(L(n)) \cong \mathbb{Z}$. The spaces $L(n)$ are examples of *lens spaces*.

2.3   Let (X, A) be a compact pair. A continuous function $r : X \longrightarrow A$ is a *retract* if $r(a) = a$ for every a in A.

   (a) Prove that there does not exist a retract from the closed unit disk B^2 to the unit circle S^1.
   (b) Prove the *Brouwer fixed point theorem*: Every continuous function $\phi : B^2 \longrightarrow B^2$ has a fixed point; i.e., a point x in B^2 such that $\phi(x) = x$.
       Hint: Assume that ϕ has no fixed point, and for each x in B^2, define $r(x)$ to be the intersection point of S^1 and the ray that starts at x and contains $\phi(x)$.

2.4   Give an example of a function that is continuous at infinity but not proper.

2.5   Suppose that (X, A) and (A, Z) are compact pairs. Prove that there exists a six-term exact sequence

$$
\begin{array}{ccccc}
K^0(X, A) & \longrightarrow & K^0(X, Z) & \longrightarrow & K^0(A, Z) \\
\uparrow{\scriptstyle\partial} & & & & \downarrow{\scriptstyle\partial} \\
K^{-1}(A, Z) & \longleftarrow & K^{-1}(X, Z) & \longleftarrow & K^{-1}(X, A),
\end{array}
$$

and write down formulas for the two connecting maps.

2.6   Prove that $K^0(X \times S^1) \cong K^0(X) \oplus K^{-1}(X) \cong K^{-1}(X \times S^1)$ for every compact Hausdorff space X.

2.7   Let X be a compact Hausdorff space. The *cone* on X is the topological space formed by taking the product space $X \times [0, 1]$ and identifying $X \times \{1\}$ to a point. Prove that $\widetilde{K}^0(X)$ and $\widetilde{K}^{-1}(X)$ are always trivial.

2.8   The *smash product* of compact pointed spaces X and Y is the pointed space $X \wedge Y$ obtained by taking $X \times Y$ and collapsing $X \vee Y$ to a point; the basepoint is the image of $X \vee Y$ in $X \wedge Y$. Prove that for $p = 0, -1$, there exist isomorphisms

$$
\widetilde{K}^p(X \times Y) \cong \widetilde{K}^p(X \wedge Y) \oplus \widetilde{K}^p(X) \oplus \widetilde{K}^p(Y).
$$

2.9   For each compact pointed space (X, x_0), define the *reduced suspension* $\widetilde{S}X$ of X to be $S^1 \wedge X$, and for $n > 1$, inductively define $\widetilde{S}^n X$ to be $\widetilde{S}(\widetilde{S}^{n-1}X)$.

(a) Prove that $\widetilde{S}^n$ defines a covariant functor from the category of pointed spaces to itself for each $n \geq 1$.

(b) For each natural number n, define $\widetilde{K}^{-n}(X)$ to be the kernel of the homomorphism from $K^{-n}(X)$ to $K^{-n}(x_0)$. Prove that there is an isomorphism $\widetilde{K}^{-n}(X) \cong \widetilde{K}^0(\widetilde{S}^n X)$.

2.10   For each compact Hausdorff space X, define the *unreduced suspension* SX of X to be the quotient space obtained by forming the product $X \times [0, 1]$, identifying $X \times \{0\}$ to one point, and identifying $X \times \{1\}$ to a second point. For $n > 1$, inductively define $S^n X$ as $S(S^{n-1}X)$. Prove that $K^{-n}(S^n X) \cong K^0(\widetilde{S}^n X)$ for every natural number n.

2.11   For each compact pair (X, A) and integer n, show that the relative group $K^{-n}(X, A)$ in Definition 2.7.20 can be alternately defined as the kernel of the homomorphism

$$
\mu_2^* : K^{-n}(\mathcal{D}(X, A)) \longrightarrow K^{-n}(X).
$$

3

Additional structure

In this chapter we will discuss various aspects of K-theory. Some of these results have to do with additional algebraic structures, while others are facts that we will need in Chapter 4.

3.1 Mayer–Vietoris

The *Mayer–Vietoris* exact sequence relates the K-theory of a compact Hausdorff space X to the K-theory of two closed subspaces of X. In addition to being a useful computational tool, the Mayer-Vietoris sequence will play an important role later in the chapter. The existence of the exact sequence is a consequence of the following homological algebra fact.

Lemma 3.1.1 *Suppose*

$$
\begin{array}{ccccccccccc}
\mathcal{G}_1 & \xrightarrow{\phi_1} & \mathcal{G}_2 & \xrightarrow{\phi_2} & \mathcal{G}_3 & \xrightarrow{\phi_3} & \mathcal{G}_4 & \xrightarrow{\phi_4} & \mathcal{G}_5 & \xrightarrow{\phi_5} & \mathcal{G}_6 \\
\downarrow{\alpha_1} & & \downarrow{\alpha_2} & & \downarrow{\alpha_3} & & \downarrow{\alpha_4} & & \downarrow{\alpha_5} & & \downarrow{\alpha_6} \\
\mathcal{H}_1 & \xrightarrow{\psi_1} & \mathcal{H}_2 & \xrightarrow{\psi_2} & \mathcal{H}_3 & \xrightarrow{\psi_3} & \mathcal{H}_4 & \xrightarrow{\psi_4} & \mathcal{H}_5 & \xrightarrow{\psi_5} & \mathcal{H}_6
\end{array}
$$

is a diagram of abelian groups that is commutative and has exact rows. Further suppose that α_1 and α_4 are isomorphisms. Then the sequence

$$
\mathcal{G}_2 \xrightarrow{(\alpha_2,\phi_2)} \mathcal{H}_2 \oplus \mathcal{G}_3 \xrightarrow{-\psi_2+\alpha_3} \mathcal{H}_3 \xrightarrow{\phi_4\alpha_4^{-1}\psi_3} \mathcal{G}_5 \xrightarrow{(\alpha_5,\phi_5)} \mathcal{H}_5 \oplus \mathcal{G}_6
$$

is exact.

Proof Take g_2 in $\mathcal{G}_2$. Then $\alpha_3\phi_2(g_2) = \psi_2\alpha_2(g_2)$ and so $-\psi_2\alpha_2(g_2) + \alpha_3\phi_2(g_2) = 0$ in $\mathcal{H}_3$. Thus the kernel of $-\psi_2 + \alpha_3$ contains the image of

(α_2, ϕ_2). To show the reverse containment, take (h_2, g_3) in $\mathcal{H}_2 \oplus \mathcal{G}_3$ and suppose that $-\psi_2(h_2) + \alpha_3(g_3) = 0$. Then

$$\alpha_4 \phi_3(g_3) = \psi_3 \alpha_3(g_3) = \psi_3 \psi_2(h_2) = 0$$

in $\mathcal{H}_4$, and $\phi_3(g_3) = 0$ in $\mathcal{G}_4$ because α_4 is an isomorphism. Thus there exists an element $\tilde{g}_2$ in $\mathcal{G}_2$ such that $\phi_2(\tilde{g}_2) = g_3$. Because

$$\psi_2 \alpha_2(\tilde{g}_2) = \alpha_3 \phi_2(\tilde{g}_2) = \alpha_3(g_3) = \psi_2(h_2),$$

we see that $\alpha_2(\tilde{g}_2) - h_2$ is in the kernel of ψ_2. Choose h_1 in $\mathcal{H}_1$ with the property that $\psi_1(h_1) = \alpha_2(\tilde{g}_2) - h_2$, let $g_1 = \alpha_1^{-1}(h_1)$, and consider the element $\tilde{g}_2 - \phi_1(g_1)$ in $\mathcal{G}_2$. Then

$$\phi_2(\tilde{g}_2 - \phi_1(g_1)) = \phi_2(\tilde{g}_2) - \phi_2 \phi_1(g_1) = \phi_2(\tilde{g}_2) = g_3$$

and

$$\alpha_2(\tilde{g}_2 - \phi_1(g_1)) = \alpha_2(\tilde{g}_2) - \alpha_2 \phi_1(g_1) = \alpha_2(\tilde{g}_2) - \psi_1 \alpha_1(g_1)$$
$$= \alpha_2(\tilde{g}_2) - \psi_1(h_1) = \alpha_2(\tilde{g}_2) - (\alpha_2(\tilde{g}_2) - h_2) = h_2,$$

whence our sequence is exact at $\mathcal{H}_2 \oplus \mathcal{G}_3$.

To show exactness at $\mathcal{H}_3$, take (h_2, g_3) in $\mathcal{H}_2 \oplus \mathcal{G}_3$. Then

$$\phi_4 \alpha_4^{-1} \psi_3(-\psi_2(h_2) + \alpha_3(g_3)) =$$
$$- \phi_4 \alpha_4^{-1} \psi_3 \psi_2(h_2) + \phi_4 \alpha_4^{-1} \psi_3 \alpha_3(g_3) = 0 + \phi_4 \phi_3(g_3) = 0.$$

Now suppose that $\phi_4 \alpha_4^{-1} \psi_3(h_3) = 0$ in $\mathcal{G}_5$. Then $\alpha_4^{-1} \psi_3(h_3) = \phi_3(\tilde{g}_3)$ for some $\tilde{g}_3$ in $\mathcal{G}_3$, and

$$\psi_3 \alpha_3(\tilde{g}_3) = \alpha_4 \phi_3(\tilde{g}_3) = \alpha_4 \alpha_4^{-1} \psi_3(h_3) = \psi_3(h_3).$$

Thus $\alpha_3(\tilde{g}_3) - h_3$ is in the kernel of ψ_3 and therefore in the image of ψ_2. Choose $\tilde{h}_2$ in $\mathcal{H}_2$ with $\psi_2(\tilde{h}_2) = \alpha_3(\tilde{g}_3) - h_3$. Then

$$-\psi_2(\tilde{h}_2) + \alpha_3(\tilde{g}_3) = -(\alpha_3(\tilde{g}_3) - h_3) + \alpha_3(\tilde{g}_3) = h_3.$$

Finally, we show exactness at $\mathcal{G}_5$. Take h_3 in $\mathcal{H}_3$, and note that

$$\left(\alpha_5 \phi_4 \alpha_4^{-1} \psi_3(h_3), \phi_5 \phi_4 \alpha_4^{-1} \psi_3(h_3)\right) = \left(\psi_4 \alpha_4 \alpha_4^{-1} \psi_3(h_3), 0\right) = (0, 0).$$

Going the other way, suppose $(\alpha_5(g_5), \phi_5(g_5)) = 0$ in $\mathcal{H}_5 \oplus \mathcal{G}_6$ for some g_5 in $\mathcal{G}_5$. Then $\phi_4(g_4) = g_5$ for some g_4 in $\mathcal{G}_4$, and

$$\psi_4 \alpha_4(g_4) = \alpha_5 \phi_4(g_4) = \alpha_5(g_5) = 0.$$

Thus we may choose $\tilde{h}_3$ in $\mathcal{H}_3$ with the property that $\psi_3(\tilde{h}_3) = \alpha_4(g_4)$, whence

$$\phi_4 \alpha_4^{-1} \psi_3(\tilde{h}_3) = \phi_4 \alpha_4^{-1} \alpha_4(g_4) = \phi_4(g_4) = g_5,$$

as desired. $\qquad\qquad\qquad\qquad\qquad\qquad\qquad\qquad\qquad\qquad\square$

Theorem 3.1.2 (Mayer–Vietoris) *Suppose that X is locally compact Hausdorff and that A_1 and A_2 are closed subspaces of X whose union is X. Then there exists an exact sequence*

$$\begin{array}{ccc}
K^0(X) \longrightarrow K^0(A_1) \oplus K^0(A_2) \longrightarrow K^0(A_1 \cap A_2) \\
\downarrow \qquad\qquad\qquad\qquad\qquad\qquad\qquad \downarrow \\
K^{-1}(A_1 \cap A_2) \longleftarrow K^{-1}(A_1) \oplus K^{-1}(A_2) \longleftarrow K^{-1}(X).
\end{array}$$

Proof For $p = 0, -1$, consider the commutative diagram

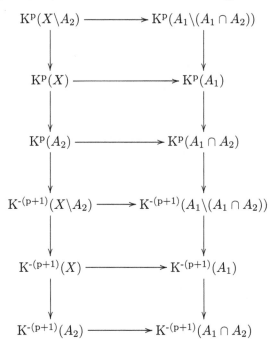

whose horizontal homomorphisms are induced by inclusion maps. The columns of the diagram are exact by Corollary 2.7.17, and because $A_1 \backslash A_1 \cap A_2 = X \backslash A_2$, the theorem immediately follows from Lemma 3.1.1. $\qquad\qquad\qquad\qquad\qquad\qquad\qquad\qquad\qquad\qquad\square$

3.2 Tensor products

Given a compact Hausdorff space X, the addition on $K^0(X)$ comes from the internal Whitney sum of vector bundles. We can also define a multiplication on vector bundles called *tensor product*; we will show that tensor product makes $K^0(X)$ into a commutative ring with unit. We begin by looking at the tensor product of *modules*.

Definition 3.2.1 *Let $\mathcal{R}$ be a ring with unit. A (left) $\mathcal{R}$-module is an abelian group $\mathcal{M}$ equipped with an* action *of $\mathcal{R}$ on $\mathcal{M}$; i.e., a function $\mathcal{R} \times \mathcal{M} \longrightarrow \mathcal{M}$, denoted by juxtaposition or a dot, such that for all r and r' in $\mathcal{R}$ and a and a' in $\mathcal{M}$:*

(i) $r(a + a') = ra + ra'$;

(ii) $(r + r')a = ra + r'a$;

(iii) $r(r'a) = (rr')a$;

(iv) $1a = a$.

A map $\phi : \mathcal{M} \longrightarrow \mathcal{M}'$ is an $\mathcal{R}$-module homomorphism if ϕ is a group homomorphism with the additional property that $\phi(ra) = r\phi(a)$ for all r in $\mathcal{R}$ and a in $\mathcal{M}$. If there exists a finite subset $\{a_1, a_2, \ldots, a_n\}$ of $\mathcal{M}$ with the feature that every element a of $\mathcal{M}$ can be written (not necessarily uniquely!) in the form $a = r_1 a_1 + r_2 a_2 + \cdots + r_n a_n$ for some $r_1, r_2, \ldots, r_n$ in $\mathcal{R}$, we say that $\mathcal{M}$ is finitely generated. *Finally, if $\mathcal{M}$ is isomorphic to $\mathcal{R}^n$ for some natural number n, we say that $\mathcal{M}$ is a free $\mathcal{R}$-module of rank n.*

Example 3.2.2 *Every abelian group is a $\mathbb{Z}$-module and an abelian group is a free $\mathbb{Z}$-module if and only if it is a free abelian group; i.e., isomorphic to a direct sum of copies of $\mathbb{Z}$.*

Example 3.2.3 *Let $\mathbb{F}$ be a field. An $\mathbb{F}$-module is a vector space over $\mathbb{F}$, and thus every $\mathbb{F}$-module is free.*

Example 3.2.4 *Let $\mathcal{R}$ be any ring and suppose $a_1, a_2, \ldots, a_m$ are objects that are not in $\mathcal{R}$. The* free module generated by $a_1, a_2, \ldots, a_m$ *is the set of formal sums*

$$\{r_1 a_1 + r_2 a_2 + \cdots + r_m a_m : r_1, r_2, \ldots, r_m \in \mathcal{R}\}$$

with addition and the action of $\mathcal{R}$ defined by the formulas

$$(r_1 a_1 + r_2 a_2 + \cdots + r_m a_m) + (r_1' a_1 + r_2' a_2 + \cdots + r_m' a_m) =$$
$$(r_1 + r_1') a_1 + (r_2 + r_2') a_2 + \cdots + (r_m + r_m') a_m$$

$$r'(r_1 a_1 + r_2 a_2 + \cdots + r_m a_m) = r' r_1 a_1 + r' r_2 a_2 + \cdots + r' r_m a_m.$$

Definition 3.2.5 *Let $\mathcal{R}$ be a* commutative *ring with unit and suppose that $\mathcal{M}$ and $\mathcal{N}$ are $\mathcal{R}$-modules. The* tensor product *of $\mathcal{M}$ and $\mathcal{N}$ (over $\mathcal{R}$) is the $\mathcal{R}$-module $\mathcal{M} \otimes \mathcal{N}$ generated by* simple tensors

$$\{a \otimes b : a \in \mathcal{M}, b \in \mathcal{N}\}$$

subject to the relations

$$(a + a') \otimes b = a \otimes b + a' \otimes b$$
$$a \otimes (b + b') = a \otimes b + a \otimes b'$$
$$r(a \otimes b) = (ra) \otimes b = a \otimes (rb)$$

for all a and a' in $\mathcal{M}$, all b and b' in $\mathcal{N}$, and all r in $\mathcal{R}$.

We could also define the tensor product of modules over a noncommutative ring, but we will not need this. In fact, the only module tensor products we will use in this book are tensor products of abelian groups ($\mathbb{Z}$-modules) and tensor products of vector spaces ($\mathbb{C}$-modules).

Example 3.2.6 *Let $\mathcal{G}$ be an abelian group. For each natural number k, there is a $\mathbb{Z}$-module isomorphism*

$$\mathcal{G} \otimes \mathbb{Z}^k \cong \bigoplus_{i=1}^{k} \mathcal{G}.$$

We can produce such an isomorphism in the following way: for each $1 \le i \le k$, let δ_i be the element of $\mathbb{Z}^n$ that is 1 is the ith position and is 0 elsewhere. Then every element of $\mathcal{G} \otimes \mathbb{Z}^n$ can be uniquely written in the form $\sum_{i=1}^{k} g_i \otimes \delta_i$, and this is mapped to the k-tuple $(g_1, g_2, \ldots, g_k)$.

Tensor products with $\mathbb{Z}^k$ have an important property that does not generally hold when tensoring with other abelian groups.

Proposition 3.2.7 *Suppose that*

$$\mathcal{G}_1 \xrightarrow{\phi_1} \mathcal{G}_2 \xrightarrow{\phi_2} \cdots \xrightarrow{\phi_{n-1}} \mathcal{G}_n$$

is an exact sequence of abelian groups. Then for every natural number n, the sequence

$$\mathcal{G}_1 \otimes \mathbb{Z}^k \xrightarrow{\phi_1 \otimes id} \mathcal{G}_2 \otimes \mathbb{Z}^k \xrightarrow{\phi_2 \otimes id} \cdots \xrightarrow{\phi_{n-1} \otimes id} \mathcal{G}_k \otimes \mathbb{Z}^k$$

is also exact.

Proof Using the isomorphism described in Example 3.2.6, the sequence in question becomes

$$\bigoplus_{i=1}^k \mathcal{G}_1 \xrightarrow{\oplus \phi_1} \bigoplus_{i=1}^k \mathcal{G}_2 \xrightarrow{\oplus \phi_2} \cdots \xrightarrow{\oplus \phi_{n-1}} \bigoplus_{i=1}^k \mathcal{G}_n,$$

which is exact.   □

Example 3.2.8 *Let V and W be vector spaces. The tensor product $V \otimes W$ (over $\mathbb{C}$) is also a vector space. If $\{e_i : 1 \leq i \leq m\}$ and $\{f_j : 1 \leq j \leq n\}$ are bases for V and W respectively, then $\{e_i \otimes f_j : 1 \leq i \leq m, 1 \leq j \leq n\}$ is a vector space basis for $V \otimes W$.*

Proposition 3.2.9 *Let V and W be inner product spaces. Then there is a unique inner product on $V \otimes W$ for which*

$$\langle v \otimes w, v' \otimes w' \rangle = \langle v, v' \rangle \langle w, w' \rangle$$

for all simple tensors $v \otimes w$ and $v' \otimes w'$.

Proof Choose orthonormal bases $\{e_i : 1 \leq i \leq m\}$ and $\{f_j : 1 \leq j \leq n\}$ for V and W respectively. Then $\{e_i \otimes f_j : 1 \leq i \leq m, 1 \leq j \leq n\}$ is a vector space basis for $V \otimes W$. Define an inner product on $V \otimes W$ by the formula

$$\left\langle \sum_{i,j} \alpha_{ij} e_i \otimes f_j, \sum_{i,j} \beta_{ij} e_i \otimes f_j \right\rangle = \sum_{i,j} \alpha_{ij} \overline{\beta}_{ij}.$$

This inner product satisfies the desired equation when applied to simple tensors and is unique because every element of $V \otimes W$ is a complex linear combination of simple tensors.   □

Definition 3.2.10 *Let V, V', W and W' be vector spaces, and suppose that $\phi : V \longrightarrow V'$ and $\psi : W \longrightarrow W'$ are linear maps. We define*

$$\phi \otimes \psi : V \otimes W \longrightarrow V' \otimes W'$$

by setting

$$(\phi \otimes \psi)(v \otimes w) = \phi(v) \otimes \psi(w)$$

on simple tensors and extending linearly.

By its very definition $\phi \otimes \psi$ is linear, and an easy check shows that $\phi \otimes \psi$ is well defined.

We can also write the tensor product of linear transformations in terms of matrices. Choose an ordered basis $\{e_1, e_2, \ldots, e_m\}$ of $\mathcal{V}$ and let A be the matrix of ϕ with respect to this basis. Similarly, let B be the matrix representation of ψ with respect to an ordered basis $\{f_1, f_2, \ldots, f_n\}$ of $\mathcal{W}$. Then

$$\{e_1 \otimes f_1, e_1 \otimes f_1, \ldots, e_1 \otimes f_n, e_2 \otimes f_1, \ldots, e_m \otimes f_{n-1}, e_m \otimes f_n\}$$

is an ordered basis for $\mathcal{V} \otimes \mathcal{W}$. The matrix for $\mathsf{A} \otimes \mathsf{B}$ with respect to this basis is the mn-by-mn matrix obtained by taking each entry of A and multiplying by the matrix for B. For example, if

$$\mathsf{A} = \begin{pmatrix} a_{11} & a_{12} \\ a_{21} & a_{22} \end{pmatrix} \quad \text{and} \quad \mathsf{B} = \begin{pmatrix} b_{11} & b_{12} & b_{13} \\ b_{21} & b_{22} & b_{23} \\ b_{31} & b_{32} & b_{33} \end{pmatrix},$$

then

$$\mathsf{A} \otimes \mathsf{B} = \begin{pmatrix} a_{11}b_{11} & a_{11}b_{12} & a_{11}b_{13} & a_{12}b_{11} & a_{12}b_{12} & a_{12}b_{13} \\ a_{11}b_{21} & a_{11}b_{22} & a_{11}b_{23} & a_{12}b_{21} & a_{12}b_{22} & a_{12}b_{23} \\ a_{11}b_{31} & a_{11}b_{32} & a_{11}b_{33} & a_{12}b_{31} & a_{12}b_{32} & a_{12}b_{33} \\ a_{21}b_{11} & a_{21}b_{12} & a_{21}b_{13} & a_{22}b_{11} & a_{22}b_{12} & a_{22}b_{13} \\ a_{21}b_{21} & a_{21}b_{22} & a_{21}b_{23} & a_{22}b_{21} & a_{22}b_{22} & a_{22}b_{23} \\ a_{21}b_{31} & a_{21}b_{32} & a_{21}b_{33} & a_{22}b_{31} & a_{22}b_{32} & a_{22}b_{33} \end{pmatrix}.$$

The tensor product of matrices is sometimes called the *Kronecker product*.

Definition 3.2.11 *Let X and Y be compact Hausdorff spaces, let m and n be natural numbers, and let A and B be elements of $\mathrm{M}(m, C(X))$ and $\mathrm{M}(n, C(Y))$ respectively. The* external tensor product $\mathsf{A} \boxtimes \mathsf{B}$ *is the matrix in $\mathrm{M}(mn, C(X \times Y))$ defined by the formula*

$$(\mathsf{A} \boxtimes \mathsf{B})(x, y) = \mathsf{A}(x) \otimes \mathsf{B}(y)$$

for all x in X and y in Y.

Proposition 3.2.12 *Suppose X and Y are compact Hausdorff spaces. For all natural numbers m and n:*

   (i) *If* E *in* $\mathrm{M}(m, C(X))$ *and* F *in* $\mathrm{M}(n, C(Y))$ *are idempotents, then* $\mathsf{E} \boxtimes \mathsf{F}$ *is an idempotent in* $\mathrm{M}(mn, C(X \times Y))$.

   (ii) *If* S *is in* $\mathrm{GL}(m, C(X))$ *and* T *is in* $\mathrm{GL}(n, C(Y))$, *then* $\mathsf{S} \boxtimes \mathsf{T}$ *is in* $\mathrm{GL}(mn, C(X \times Y))$.

Proof Compute.   □

Proposition 3.2.13 *Let X and Y be compact Hausdorff spaces. Then external tensor product defines a map from* $\mathrm{Idem}(C(X)) \times \mathrm{Idem}(C(Y))$ *to* $\mathrm{Idem}(C(X \times Y))$.

Proof Suppose E in $\mathrm{M}(m, C(X))$ and F in $\mathrm{M}(n, C(Y))$ are idempotents. Then

$$\mathrm{diag}(\mathsf{E}, 0_k) \boxtimes \mathrm{diag}(\mathsf{F}, 0_l) = \mathrm{diag}(\mathsf{E} \boxtimes \mathsf{F}, \mathsf{E} \boxtimes 0_l, 0_k \boxtimes \mathsf{F}, 0_k \boxtimes 0_l)$$
$$= \mathrm{diag}(\mathsf{E} \boxtimes \mathsf{F}, 0_{kl}, 0_{kl}, 0_{kl})$$

for any two natural numbers k and l. Therefore $[\mathsf{E}\boxtimes\mathsf{F}]$ in $\mathrm{Idem}(C(X \times Y))$ is independent of the sizes of matrices used to represent E and F in $\mathrm{Idem}(C(X))$ and $\mathrm{Idem}(C(Y))$. Next, choose S in $\mathrm{GL}(m, C(X))$ and T in $\mathrm{GL}(n, C(Y))$. Definition 3.2.10 implies that

$$(\mathsf{SES}^{-1}) \boxtimes (\mathsf{TFT}^{-1}) = (\mathsf{S} \boxtimes \mathsf{T})(\mathsf{E} \boxtimes \mathsf{F})(\mathsf{S}^{-1} \boxtimes \mathsf{T}^{-1})$$
$$= (\mathsf{S} \boxtimes \mathsf{T})(\mathsf{E} \boxtimes \mathsf{F})(\mathsf{S} \boxtimes \mathsf{T})^{-1},$$

and thus $[\mathsf{E}\boxtimes\mathsf{F}]$ does not depend on the similarity classes of E or F.   □

Construction 3.2.14 *Let V and W be vector bundles over compact Hausdorff spaces X and Y respectively, and write $V = \mathrm{Ran}\,\mathsf{E}$ and $W = \mathrm{Ran}\,\mathsf{F}$ for idempotents E over X and F over Y. The external tensor product of V and W is the vector bundle $V \boxtimes W = \mathrm{Ran}(\mathsf{E}\boxtimes\mathsf{F})$ over $X \times Y$. If V and W are Hermitian vector bundles, we define a Hermitian metric on $V \boxtimes W$ by employing the inner product formula from Proposition 3.2.9 on each fiber.*

Propositions 1.7.6 and 3.2.13 guarantee that $V \boxtimes W$ is well defined up to isomorphism. For each point (x, y) in $X \times Y$, the fiber $(V \boxtimes W)_{(x,y)}$ is the vector space $V_x \otimes W_y$.

Construction 3.2.15 *Let V and W be vector bundles over the same compact Hausdorff space X and let $\Delta : X \longrightarrow X \times X$ be the diagonal map. The internal tensor product of V and W is the vector bundle $V \otimes W = \Delta^*(V \boxtimes W)$. As in Construction 3.2.14, Hermitian metrics on V and W determine a Hermitian metric on $V \otimes W$.*

3.3 Multiplicative structures

Proposition 3.3.1 *Let X and Y be compact Hausdorff spaces. The external tensor product determines a group homomorphism*

$$\mathrm{K}^0(X) \otimes \mathrm{K}^0(Y) \longrightarrow \mathrm{K}^0(X \times Y).$$

Proof Given elements $[V] - [V']$ and $[W] - [W']$ of $\mathrm{K}^0(X)$ and $\mathrm{K}^0(Y)$ respectively, we define the product on simple tensors to be

$$\left([V] - [V']\right) \boxtimes \left([W] - [W']\right) = [V \boxtimes W] - [V \boxtimes W'] - [V' \boxtimes W] + [V' \boxtimes W']$$

and then extend by the distributive law.

Suppose that $[V] - [V'] = [\widetilde{V}] - [\widetilde{V}']$ in $\mathrm{K}^0(X)$. Definition 1.6.6 states that

$$V \oplus \widetilde{V}' \oplus V'' \cong \widetilde{V} \oplus V' \oplus V''$$

for some vector bundle V'' over X. External product distributes across internal Whitney sum, whence

$$(V \boxtimes W) \oplus (\widetilde{V}' \boxtimes W) \oplus (V'' \boxtimes W) \cong (\widetilde{V} \boxtimes W) \oplus (V' \boxtimes W) \oplus (V'' \boxtimes W).$$

Therefore

$$[V \boxtimes W] - [V' \boxtimes W] = [\widetilde{V} \boxtimes W] - [\widetilde{V}' \boxtimes W],$$

and replacing W with $\widetilde{W}$ yields

$$[V \boxtimes \widetilde{W}] - [V' \boxtimes \widetilde{W}] = [\widetilde{V} \boxtimes \widetilde{W}] - [\widetilde{V}' \boxtimes \widetilde{W}].$$

Therefore the formula for the product does not depend on the choice of representative in $\mathrm{K}^0(X)$. A similar argument shows that the product is independent of our choice of representative in $\mathrm{K}^0(Y)$. Finally, an easy check shows that the defining relations of the tensor product are respected, so the homomorphism is well defined. □

If X and Y are the same space, we can say more.

Theorem 3.3.2 *Let X be compact Hausdorff. Then $K^0(X)$ is a commutative ring with unit under internal tensor product.*

Proof The tensor product of vector spaces is commutative and associative and distributes across internal Whitney sum. Propositions 3.2.13 and 3.3.1 then imply that $K^0(X)$ is a commutative ring with $[\Theta^1(X)]$ as the multiplicative identity. $\qquad\square$

Proposition 3.3.3 *The assignment $X \mapsto K^0(X)$ determines a contravariant functor K^0 from the category of compact Hausdorff spaces to the category of commutative rings with unital ring homomorphisms as the morphisms.*

Proof We know from Proposition 2.1.6 that K^0 is a functor into the category of abelian groups. Let X and Y be compact Hausdorff spaces and suppose that $\phi : X \longrightarrow Y$ is continuous. Then a computation shows that $\phi^* : M(n, C(Y)) \longrightarrow M(n, C(X))$ is a unital ring homomorphism for every natural number n, whence the desired result follows. $\qquad\square$

Example 3.3.4 *Let X be any compact Hausdorff space with the property that $K^0(X) \cong \mathbb{Z}$ as groups. For all natural numbers k and l, the definition of tensor product immediately yields $[I_k] \otimes [I_l] = I_{kl}$, and thus $K^0(X)$ and $\mathbb{Z}$ are isomorphic rings.*

To compute the ring structure of $K^0(X)$ in some more interesting cases, we extend the definition of multiplication in K^0 to locally compact Hausdorff spaces. First we record a result we will need later and that is interesting in its own right.

Proposition 3.3.5 *For every compact Hausdorff space X, the subset of $K^0(X)$ consisting of isomorphism classes of line bundles forms an abelian group under internal tensor product.*

Proof The only nonobvious point to check is the existence of multiplicative inverses. For each line bundle V, let V^* denote its dual bundle, and define $\gamma : V^* \otimes V \longrightarrow \Theta^1(X)$ by setting $\gamma(\phi \otimes v) = \phi(v)$ and extending linearly. Then γ is a bundle isomorphism from $V^* \otimes V$ to $\Theta^1(X)$. $\qquad\square$

Theorem 3.3.6 *For every pair X, Y of locally compact Hausdorff spaces, there exists a unique group homomorphism*

$$K^0(X) \otimes K^0(Y) \longrightarrow K^0(X \times Y)$$

with the following property: let i_X and i_Y be the inclusions of $K^0(X)$ and $K^0(Y)$ into $K^0(X^+)$ and $K^0(Y^+)$ respectively, and let ι be the homomorphism in K-theory induced by the inclusion of the open subspace $X \times Y$ into the compact Hausdorff space $X^+ \times Y^+$. Then the diagram

$$
\begin{array}{ccc}
K^0(X) \otimes K^0(Y) & \longrightarrow & K^0(X \times Y) \\
\Big\downarrow{\scriptstyle i_X \otimes i_Y} & & \Big\downarrow{\scriptstyle \iota} \\
K^0(X^+) \otimes K^0(Y^+) & \xrightarrow{\ m\ } & K^0(X^+ \times Y^+)
\end{array}
$$

commutes.

Proof The image of the map $i_X \times i_Y$ contains all points of $X^+ \times Y^+$ except for those of the form (x, ∞_Y) for some x in X^+ or (∞_X, y) for some y in Y^+. Thus the complement of $(i_X \times i_Y)(X \times Y)$ in $X^+ \times Y^+$ is a subspace homeomorphic to $X^+ \vee Y^+$; the basepoint ∞ of $X^+ \vee Y^+$ is (∞_X, ∞_Y). If j denotes the inclusion of $X^+ \vee Y^+$ into $X^+ \times Y^+$, then Corollary 2.7.17 yields the exact sequence

$$K^{-1}(X^+ \times Y^+) \xrightarrow{\ j^*\ } K^{-1}(X^+ \vee Y^+) \xrightarrow{\ \partial\ }$$

$$K^0(X \times Y) \longrightarrow K^0(X^+ \times Y^+) \xrightarrow{\ j^*\ } K^0(X^+ \vee Y^+).$$

Suppose that S is an element of $GL(n, C(X^+ \vee Y^+))$ for some natural number n. Let S_{X^+} and S_{Y^+} be the restrictions of S to $GL(n, C(X^+))$ and $GL(n, C(Y^+))$ respectively and define $\widetilde{S}$ in $GL(n, C(X^+ \times Y^+))$ by the formula

$$\widetilde{S}(x, y) = S_{X^+}(x) S_{X^+}(\infty) S_{Y^+}(y)$$

for all (x, y) in $X^+ \times Y^+$. Then $\widetilde{S}$ restricts to S and therefore the homomorphism $j^* : K^{-1}(X^+ \times Y^+) \longrightarrow K^{-1}(X^+ \vee Y^+)$ is surjective. Thus the connecting map ∂ is zero and we obtain the short exact sequence

$$0 \longrightarrow K^0(X \times Y) \longrightarrow K^0(X^+ \times Y^+) \xrightarrow{\ j^*\ } K^0(X^+ \vee Y^+).$$

Let $\psi_Y : X^+ \vee Y^+ \longrightarrow Y^+$ be the function that collapses all of X to the basepoint of $X^+ \vee Y^+$ and similarly define $\psi_X : X^+ \vee Y^+ \longrightarrow X^+$. Theorem 2.2.10 gives us an isomorphism

$$\mu_1^* + \psi_Y^* : \mathrm{K}^0(X^+ \vee Y^+, Y^+) \oplus \mathrm{K}^0(Y^+) \longrightarrow \mathrm{K}^0(X^+ \vee Y^+)$$

that maps the subgroup of $\mathrm{K}^0(Y^+)$ generated by trivial bundles to the subgroup of $\mathrm{K}^0(X^+ \vee Y^+)$ generated by trivial bundles. We also have an isomorphism $q^* : \widetilde{\mathrm{K}}^0(X^+) \longrightarrow \mathrm{K}^0(X^+ \vee Y^+, Y^+)$ from Theorem 2.2.14 and Proposition 2.5.3, and these combine to give us the isomorphism

$$\psi_X^* + \psi_Y^* : \widetilde{\mathrm{K}}^0(X^+) \oplus \widetilde{\mathrm{K}}^0(Y^+) \longrightarrow \widetilde{\mathrm{K}}^0(X^+ \vee Y^+).$$

If j_X and j_Y denote the inclusions of X^+ and Y^+ respectively into $X^+ \vee Y^+$, then $(j_X^* \oplus j_Y^*)(\psi_X^* + \psi_Y^*)$ is the identity on $\widetilde{\mathrm{K}}^0(X^+) \oplus \widetilde{\mathrm{K}}^0(Y^+)$ and hence

$$j_X^* \oplus j_Y^* : \widetilde{\mathrm{K}}^0(X^+ \vee Y^+) \longrightarrow \widetilde{\mathrm{K}}^0(X^+) \oplus \widetilde{\mathrm{K}}^0(Y^+)$$

is also an isomorphism.

Take elements $[\mathsf{E}] - [I_k]$ and $[\mathsf{F}] - [I_l]$ in $\mathrm{K}^0(X) = \widetilde{\mathrm{K}}^0(X^+)$ and $\mathrm{K}^0(Y) = \widetilde{\mathrm{K}}^0(Y^+)$ respectively; we may assume by Proposition 2.5.6 that $\mathsf{E}(\infty_X) = I_k$ and that $\mathsf{F}(\infty_Y) = I_l$. Then $([\mathsf{E}] - [I_k]) \boxtimes ([\mathsf{F}] - [I_l])$ is in $\mathrm{K}^0(X^+ \times Y^+)$ and $j^*(([\mathsf{E}] - [I_k]) \boxtimes ([\mathsf{F}] - [I_l]))$ is an element of $\widetilde{\mathrm{K}}^0(X^+ \vee Y^+)$, and

$$j_X^* \Big(j^*(([\mathsf{E}] - [I_k]) \boxtimes ([\mathsf{F}] - [I_l])) \Big) = ([\mathsf{E}|X^+] - [I_k]) \boxtimes ([I_l] - [I_l]) = 0$$

$$j_Y^* \Big(j^*(([\mathsf{E}] - [I_k]) \boxtimes ([\mathsf{F}] - [I_l])) \Big) = ([I_k] - [I_k]) \boxtimes ([\mathsf{F}|Y^+] - [I_l]) = 0.$$

Therefore the image of $([\mathsf{E}] - [I_k]) \boxtimes ([\mathsf{F}] - [I_l])$ in $\mathrm{K}^0(X^+ \vee Y^+)$ is zero, whence $([\mathsf{E}] - [I_k]) \boxtimes ([\mathsf{F}] - [I_l])$ is an element of $\mathrm{K}^0(X \times Y)$, as desired. The diagram in the statement of the proposition commutes by construction and uniqueness of the homomorphism follows from the injectivity of ι. $\qquad\square$

Corollary 3.3.7 *Let (X, A) and (Y, B) be compact pairs. Then there exists a unique group homomorphism*

$$\mathrm{K}^0(X, A) \otimes \mathrm{K}^0(Y, B) \longrightarrow \mathrm{K}^0\big(X \times Y, (X \times B) \cup (A \times Y)\big)$$

such that the diagram

$$
\begin{array}{ccc}
\mathrm{K}^0(X, A) \times \mathrm{K}^0(Y, B) & \longrightarrow & \mathrm{K}^0\big(X \times Y, (X \times B) \cup (A \times Y)\big) \\
{\scriptstyle \mu_1^* \otimes \mu_1^*}\Big\downarrow & & \Big\downarrow{\scriptstyle \mu_1^*} \\
\mathrm{K}^0(X) \times \widetilde{\mathrm{K}}^0(Y) & \xrightarrow{\quad \boxtimes \quad} & \mathrm{K}^0(X \times Y)
\end{array}
$$

commutes.

Proof The existence and uniqueness of the desired homomorphism follow immediately from Theorem 2.2.14, Proposition 2.5.3, Theorem 3.3.6, and the homeomorphism

$$(X \times Y)\backslash((X \times B) \cup (A \times Y)) \cong (X\backslash A) \times (Y\backslash B).$$

An easy check verifies the commutativity of the diagram. □

Corollary 3.3.8 *Let X be compact Hausdorff and suppose that A and A' are closed subspaces of X. Then there exists a unique group homomorphism*

$$K^0(X, A) \otimes K^0(X, A') \longrightarrow K^0(X, A \cup A')$$

that makes the diagram

$$
\begin{array}{ccc}
K^0(X, A) \otimes K^0(Y, A') & \longrightarrow & K^0(X, A \cup A') \\
{\scriptstyle \mu_1^* \otimes \mu_1^*} \downarrow & & \downarrow {\scriptstyle \mu_1^*} \\
K^0(X) \otimes K^0(X) & \xrightarrow{\ \otimes\ } & K^0(X)
\end{array}
$$

commute.

Proof Take $X = Y$ in Corollary 3.3.7 and compose external multiplication with the diagonal map

$$\Delta : (X, A \cup A') \longrightarrow (X \times X, (X \cup A') \cup (A \cup X)).$$

□

Before leaving this section, we use external multiplication to define some isomorphisms that are closely related to the Bott periodicity isomorphism.

Proposition 3.3.9 *For every compact Hausdorff space X, external multiplication by*

$$\tilde{\mathbf{b}} = \left[\left(\begin{array}{cc} |z|^2 & z\sqrt{1 - |z|^2} \\ \bar{z}\sqrt{1 - |z|^2} & 1 - |z|^2 \end{array} \right), \mathrm{diag}(1, 0) \right] - [1, 1]$$

in $K^0(B^2, S^1)$ defines an isomorphism $K^0(X) \cong K^0(X \times B^2, X \times S^1)$.

Proof If E is an idempotent in $M(n, C(X))$, then the Bott periodicity isomorphism maps $[E]$ to $[I_n - E + Ez]$ in $K^{-1}(S'X) \cong K^{-1}(X \times (S^1 \backslash \{1\}))$.

Let j be the inclusion of X into $X \times B^2$ as $X \times \{1\}$ and let $\psi : X \times B^2 \longrightarrow X$ be the projection map onto the first factor. Then ψj is the identity on X, and the contractibility of B^2 implies that $j\psi$ is homotopic to the identity map on $X \times B^2$. Therefore $j^* : K^P(X \times B^2) \longrightarrow K^P(X)$ is an isomorphism for $p = 0, -1$, and Theorems 2.2.10 and 2.3.15 imply that the K-theory of $X \times (B^2 \backslash \{1\})$ is trivial. Thus an application of Corollary 2.7.19 to the compact pair $(X \times (B^2 \backslash \{1\}), X \times (S^1 \backslash \{1\}))$ yields an isomorphism

$$\partial : K^{-1}(X \times (S^1 \backslash \{1\})) \longrightarrow K^0(X \times (B^2 \backslash \{1\}), X \times (S^1 \backslash \{1\})).$$

The matrix

$$T = \begin{pmatrix} I_n - E + Ez & -E\sqrt{1 - |z|^2} \\ E\sqrt{1 - |z|^2} & I_n - E + Ez \end{pmatrix}$$

in $GL(2n, C(X \times (B^2 \backslash \{1\})))$ restricts to $\mathrm{diag}(I_n - E + Ez, I_n - E + Ez^{-1})$ on $X \times (S^1 \backslash \{1\})$, and the formula for the connecting map ∂ in Theorem 2.6.16 gives us

$$\partial[I_n - E + Ez] = [T \, \mathrm{diag}(I_n, 0_n) T^{-1}, \mathrm{diag}(I_n, 0_n)] - [I_n, I_n]$$
$$= \left[\begin{pmatrix} I_n - E + E|z|^2 & Ez\sqrt{1 - |z|^2} \\ E\bar{z}\sqrt{1 - |z|^2} & E(1 - |z|^2) \end{pmatrix}, \mathrm{diag}(I_n, 0_n) \right] - [I_n, I_n].$$

By Corollary 2.2.15, there exists a group isomorphism

$$K^0(X \times (B^2 \backslash \{1\}), X \times (S^1 \backslash \{1\})) \cong K^0(X \times B^2, X \times S^1)),$$

and the expression for $\partial[I_n - E + Ez]$ does not change under this isomorphism. Therefore the map

$$[E] \mapsto \left[\begin{pmatrix} I_n - E + E|z|^2 & Ez\sqrt{1 - |z|^2} \\ E\bar{z}\sqrt{1 - |z|^2} & E(1 - |z|^2) \end{pmatrix}, \mathrm{diag}(I_n, 0_n) \right] - [I_n, I_n]$$

defines a isomorphism from $K^0(X)$ to $K^0(X \times B^2, X \times S^1)$. On the other hand, from Lemma 1.7.3 we have

$$[\mathsf{E}] \boxtimes \tilde{\mathbf{b}} = \left[\begin{pmatrix} \mathsf{E}|z|^2 & \mathsf{E}z\sqrt{1-|z|^2} \\ \mathsf{E}\bar{z}\sqrt{1-|z|^2} & \mathsf{E}(1-|z|^2) \end{pmatrix}, \operatorname{diag}(I_n, 0_n)\right] - [\mathsf{E}, \mathsf{E}]$$

$$= \left[\begin{pmatrix} \mathsf{E}|z|^2 & \mathsf{E}z\sqrt{1-|z|^2} \\ \mathsf{E}\bar{z}\sqrt{1-|z|^2} & \mathsf{E}(1-|z|^2) \end{pmatrix}, \operatorname{diag}(I_n, 0_n)\right] +$$

$$[\operatorname{diag}(I_n - \mathsf{E}, 0_n), \operatorname{diag}(I_n - \mathsf{E}, 0_n)] - \Big([\mathsf{E}, \mathsf{E}] + [I_n - \mathsf{E}, I_n - \mathsf{E}]\Big)$$

$$= \left[\begin{pmatrix} I_n - \mathsf{E} + \mathsf{E}|z|^2 & \mathsf{E}z\sqrt{1-|z|^2} \\ \mathsf{E}\bar{z}\sqrt{1-|z|^2} & \mathsf{E}(1-|z|^2) \end{pmatrix}, \operatorname{diag}(I_n, 0_n)\right] - [I_n, I_n],$$

whence the proposition follows. $\qquad\square$

Corollary 3.3.10 *Let (X, A) be a compact pair and let $\tilde{\mathbf{b}}$ be the element of $\mathrm{K}^0(B^2, S^1)$ defined in Proposition 3.3.9. Then external multiplication by $\tilde{\mathbf{b}}$ defines an isomorphism*

$$\mathrm{K}^0(X, A) \cong \mathrm{K}^0\big(X \times B^2, (X \times S^1) \cup (A \times B^2)\big).$$

Proof Consider the commutative diagram

$$
\begin{array}{ccc}
0 & & 0 \\
\downarrow & & \downarrow \\
\mathrm{K}^0(X, A) & & \mathrm{K}^0\big(X \times B^2, (X \times S^1) \cup (A \times B^2)\big) \\
\downarrow & & \downarrow \\
\mathrm{K}^0\big((X\backslash A)^+\big) & \xrightarrow{\boxtimes\tilde{\mathbf{b}}} & \mathrm{K}^0\big((X\backslash A)^+ \times B^2, (X\backslash A)^+ \times S^1\big) \\
\downarrow & & \downarrow \\
\mathrm{K}^0(\{\infty\}) & \xrightarrow{\boxtimes\tilde{\mathbf{b}}} & \mathrm{K}^0\big(\{\infty\} \times B^2, \{\infty\} \times S^1\big).
\end{array}
$$

The first column is exact by Theorem 2.2.14 and Proposition 2.5.3. Note that the homomorphism

$$j^* : \mathrm{K}^0\big((X\backslash A)^+ \times (B^2\backslash S^1)\big) \longrightarrow \mathrm{K}^0\big(\{\infty\} \times (B^2\backslash S^1)\big)$$

induced by inclusion has a splitting. Therefore Theorem 2.6.11 and the isomorphisms

$$K^0((X\backslash A)\times(B^2\backslash S^1))\cong K^0(X\times B^2,(X\times S^1)\cup(A\times B^2))$$
$$K^0((X\backslash A)^+\times(B^2\backslash S^1))\cong K^0((X\backslash A)^+\times B^2,(X\backslash A)^+\times S^1)$$
$$K^0(\{\infty\}\times(B^2\backslash S^1))\cong K^0(\{\infty\}\times B^2,\{\infty\}\times S^1)$$

from Theorem 2.2.14 imply that the second column is exact as well. The corollary then follows from Proposition 1.8.9. □

Corollary 3.3.11 *Let* **b** *be the element of* $K^0(\mathbb{R}^2)$ *from Example 2.8.1. Then for every locally compact Hausdorff space* X, *external multiplication by* **b** *defines an isomorphism from* $K^0(X)$ *to* $K^0(X\times\mathbb{R}^2)$.

Proof If we identify $K^0(B^2,S^1)$ with $K^0(B^2\backslash S^1)$, then the isomorphism ϕ^* from Example 2.8.1 maps $\tilde{\mathbf{b}}$ to **b**. Therefore the corollary for X compact follows directly from Proposition 3.3.9. If X is locally compact but not compact, Proposition 1.8.9 and the commutativity of the diagram

$$
\begin{array}{ccccccc}
0 & \longrightarrow & K^0(X) & \longrightarrow & K^0(X^+) & \longrightarrow & K^0(\{\infty\}) \\
& & & & \Big\downarrow{\boxtimes\mathbf{b}} & & \Big\downarrow{\boxtimes\mathbf{b}} \\
0 & \longrightarrow & K^0(X\times\mathbb{R}^2) & \longrightarrow & K^0(X^+\times\mathbb{R}^2) & \longrightarrow & K^0(\{\infty\}\times\mathbb{R}^2)
\end{array}
$$

yield the desired result. □

Corollary 3.3.12 *For every natural number* n, *the group* $K^0(\mathbb{R}^{2n})$ *is generated by the element* $\mathbf{b}^n$.

Proof Take X to be a point in Corollary 3.3.11 and induct on n. □

We now compute a nontrivial example of the ring structure on K^0.

Example 3.3.13 *We can use Corollary 3.3.8 to compute the ring structure of* $K^0(S^n)$ *when* n *is even (Example 3.3.4 tells us that* $K^0(S^n)\cong\mathbb{Z}$ *when* n *is odd). Write* $n=2m$, *and for each closed subset* A *of* S^{2m}, *let* $\kappa(A)$ *be the kernel of the homomorphism* $K^0(S^{2m})\longrightarrow K^0(A)$ *induced by inclusion. If we denote the product defined in Corollary 3.3.8 by a dot, then that corollary implies that* $\kappa(A)\cdot\kappa(B)\subseteq\kappa(A\cup B)$ *for any two closed subsets* A *and* B *of* S^{2m}.

View S^{2m} *as the one-point compactification of* $\mathbb{R}^{2m}$. *Example 2.8.1 and Corollary 3.3.12 imply that* $K^0(S^{2m})$ *is generated as a group by* $[\Theta^1(S^{2m})]$ *and* $\mathbf{b}^m$. *We know from Example 3.3.4 that* $[\Theta^1(S^{2m})]$ *generates a subring of* $K^0(S^{2m})$ *isomorphic to* $\mathbb{Z}$. *To determine the rest of the ring structure of* $K^0(S^{2m})$, *let* N *and* S *be the closed northern and southern hemispheres of* S^{2m}. *Both* N *and* S *are contractible, and Corollary 2.1.8 implies that* $K^0(N)$ *and* $K^0(S)$ *are each generated by trivial bundles. Thus the element* $\mathbf{b}^m$ *of* $K^0(S^{2m})$ *restricts to elements of both* $\kappa(N)$ *and* $\kappa(S)$. *But*

$$\kappa(N) \cdot \kappa(S) \subseteq \kappa(N \cup S) = \kappa(S^{2m}) = 0,$$

whence $\mathbf{b}^m \cdot \mathbf{b}^m = 0$.

To give a succinct description of the ring $K^0(S^{2m})$, *let* $\mathbb{Z}[t]$ *denote the ring of polynomials in the indeterminate* t *that have integer coefficients, and let* $\langle t^2 - 1 \rangle$ *be the ideal of* $\mathbb{Z}[t]$ *generated by* $t^2 - 1$. *Then we have a ring isomorphism*

$$K^0(S^{2m}) \cong \frac{\mathbb{Z}[t]}{\langle t^2 - 1 \rangle}.$$

We end this section by showing how the external tensor product can be used to compute the K-theory groups of a product space, at least when the spaces are sufficiently nice.

Definition 3.3.14 *Let* X *be a locally compact Hausdorff space. A collection* $\mathcal{A}$ *of subsets of* X *is called a* closed adapted cover *if it has the following properties:*

(i) *each set in* $\mathcal{A}$ *is closed;*

(ii) *the union of the sets in* $\mathcal{A}$ *is* X;

(iii) *the intersection of any finite number of sets in* $\mathcal{A}$ *is either empty or the product of a compact contractible space with some finite number of copies of* $\mathbb{R}$.

We say that X *is of* finite type *if* X *admits a closed adapted cover.*

Obviously $\mathbb{R}^n$ is of finite type for every natural number n. Compact smooth manifolds (see Chapter 4) are of finite type, and the product of spaces of finite type is also of finite type.

Proposition 3.3.15 (Künneth theorem) *Let* X *be a locally compact Hausdorff space of finite type and suppose* Y *is any locally compact Hausdorff space such that* $K^0(Y)$ *and* $K^{-1}(Y)$ *are free abelian groups.*

Then there exist natural group isomorphisms

$$K^0(X \times Y) \cong (K^0(X) \otimes K^0(Y)) \oplus (K^{-1}(X) \otimes K^{-1}(Y))$$
$$K^{-1}(X \times Y) \cong (K^0(X) \otimes K^{-1}(Y)) \oplus (K^0(X) \otimes K^{-1}(Y)).$$

Proof We prove the first isomorphism; the second isomorphism follows immediately from the first one and Theorem 2.6.13. Let $\gamma_0 : K^0(X) \otimes K^0(Y) \longrightarrow K^0(X \times Y)$ be the external multiplication map, and define $\gamma_{-1} : K^{-1}(X) \otimes K^{-1}(Y) \longrightarrow K^0(X \times Y)$ to be the composition

$$K^{-1}(X) \otimes K^{-1}(Y) \cong K^0(X \times \mathbb{R}) \otimes K^0(Y \times \mathbb{R}) \longrightarrow$$
$$K^0(X \times \mathbb{R} \times Y \times \mathbb{R}) \longrightarrow K^0(X \times Y \times \mathbb{R}^2) \cong K^0(X \times Y);$$

the first isomorphism comes from Theorem 2.6.13, the second map is γ_0 applied to $X \times \mathbb{R}$ and $Y \times \mathbb{R}$, the next isomorphism is the one induced by swapping the last two factors, and the final isomorphism is the inverse of external multiplication by the element $\mathbf{b}$ from Corollary 3.3.11. We will prove that $\gamma_0 \oplus \gamma_{-1}$ is an isomorphism by using induction on the number of sets in a closed adapted cover of X.

Suppose that $X = Z \times \mathbb{R}^{2k}$ for some compact contractible space Z and some natural number k, and let Y be any locally compact Hausdorff space. The contractibility of Z implies that we have isomorphisms $K^0(Z \times \mathbb{R}^{2k}) \cong K^0(\mathbb{R}^{2k})$ and $K^0(Z \times \mathbb{R}^{2k} \times Y) \cong K^0(\mathbb{R}^{2k} \times Y)$, and k applications of Corollary 3.3.11 yield that γ_0 is an isomorphism. On the other hand, Example 2.8.1 and Theorem 2.6.13 give us

$$K^{-1}(Z \times \mathbb{R}^{2k}) \cong K^0(Z \times \mathbb{R}^{2k+1}) \cong K^0(\mathbb{R}^{2k+1}) = 0,$$

whence γ_{-1} is the zero map. Therefore $\gamma_0 \oplus \gamma_{-1}$ is an isomorphism.

Now suppose that the theorem is true for n. Choose a closed adapted cover $\mathcal{A} = \{A_1, A_2, \ldots, A_n, A_{n+1}\}$ of X and set $A = A_2 \cup A_3 \cup \cdots \cup A_n \cup A_{n+1}$. Proposition 3.2.7 and Theorems 2.6.13, 2.7.15, and 3.1.2 imply that we have an exact sequence

$$(K^0(A_1) \oplus K^0(A)) \otimes K^0(Y) \longrightarrow K^0(A_1 \cap A) \otimes K^0(Y) \longrightarrow$$

$$K^0(X) \otimes K^0(Y) \overset{\partial}{\longrightarrow} (K^0(A_1 \times \mathbb{R}) \oplus K^0(A \times \mathbb{R})) \otimes K^0(Y)$$

$$\longrightarrow K^0((A_1 \cap A) \times \mathbb{R}) \otimes K^0(Y).$$

Similarly, the sequence

$$\left(K^{-1}(A_1) \oplus K^{-1}(A)\right) \otimes K^{-1}(Y) \longrightarrow K^{-1}(A_1 \cap A) \otimes K^{-1}(Y) \longrightarrow$$

$$K^{-1}(X) \otimes K^{-1}(Y) \xrightarrow{\ \partial\ } \left(K^{-1}(A_1 \times \mathbb{R}) \oplus K^{-1}(A \times \mathbb{R})\right) \otimes K^{-1}(Y)$$

$$\longrightarrow K^{-1}\big((A_1 \cap A) \times \mathbb{R}\big) \otimes K^{-1}(Y)$$

is exact. For each closed subspace $\widetilde{A}$ of X, set

$$K^{\#}(\widetilde{A}) \otimes K^{\#}(Y) = (K^0(\widetilde{A}) \otimes K^0(Y)) \oplus (K^{-1}(\widetilde{A}) \otimes K^{-1}(Y)).$$

Then Theorem 3.1.2 and the exactness of the two sequences above imply that the diagram

$$
\begin{bmatrix} (K^{\#}(A_1) \oplus K^{\#}(A)) \\ \otimes \\ K^{\#}(Y) \end{bmatrix}
\xrightarrow{\ \gamma_0 \oplus \gamma_{-1}\ }
\begin{bmatrix} K^0(A_1 \times Y) \\ \oplus \\ K^0(A \times Y) \end{bmatrix}
$$

$$\downarrow \qquad\qquad\qquad \downarrow$$

$$K^{\#}(A_1 \cap A) \otimes K^{\#}(Y) \xrightarrow{\ \gamma_0 \oplus \gamma_{-1}\ } K^0\big((A_1 \cap A) \times Y\big)$$

$$\downarrow \qquad\qquad\qquad \downarrow$$

$$K^{\#}(X) \otimes K^{\#}(Y) \xrightarrow{\ \gamma_0 \oplus \gamma_{-1}\ } K^0(X \times Y)$$

$$\downarrow \qquad\qquad\qquad \downarrow$$

$$
\begin{bmatrix} (K^{\#}(A_1 \times \mathbb{R}) \oplus K^{\#}(A \times \mathbb{R})) \\ \otimes \\ K^{\#}(Y) \end{bmatrix}
\xrightarrow{\ \gamma_0 \oplus \gamma_{-1}\ }
\begin{bmatrix} K^0(A_1 \times \mathbb{R} \times Y) \\ \oplus \\ K^0(A \times \mathbb{R} \times Y) \end{bmatrix}
$$

$$\downarrow \qquad\qquad\qquad \downarrow$$

$$K^{\#}(A_1 \cap A) \otimes K^{\#}(Y) \xrightarrow{\ \gamma_0 \oplus \gamma_{-1}\ } K^0\big((A_1 \cap A) \times Y\big)$$

$$\downarrow \qquad\qquad\qquad \downarrow$$

$$K^{\#}(X) \otimes K^{\#}(Y) \xrightarrow{\ \gamma_0 \oplus \gamma_{-1}\ } K^0(X \times Y)$$

has exact columns, and the commutativy of the diagram is due to the naturality of all the maps involved. The desired result follows from Proposition 1.8.10 and the principle of mathematical induction. □

3.4 An alternate picture of relative K^0

In this section we consider another way to define the relative version of K^0. This alternate picture is especially helpful when considering products and will play a crucial role in our proof of the Thom isomorphism theorem later in the chapter.

Definition 3.4.1 *For every compact pair (X, A), define the set*

$$\mathfrak{V}(X, A) = \{(V, W, \sigma) : V \text{ and } W \text{ are vector bundles over } X$$
$$\text{and } \sigma \text{ is a bundle isomorphism from } V|A \text{ to } W|A\}.$$

Triples (V, W, σ) and $(\widetilde{V}, \widetilde{W}, \widetilde{\sigma})$ in $\mathcal{V}(X, A)$ are isomorphic if there exist bundle isomorphisms $\nu : V \longrightarrow \widetilde{V}$ and $\omega : W \longrightarrow \widetilde{W}$ with the property that the diagram

$$
\begin{array}{ccc}
V|A & \xrightarrow{\ \sigma\ } & W|A \\
{\scriptstyle \nu|A}\downarrow & & \downarrow{\scriptstyle \omega|A} \\
\widetilde{V}|A & \xrightarrow{\ \widetilde{\sigma}\ } & \widetilde{W}|A
\end{array}
$$

commutes. In this case, we say that ν and ω implement the isomorphism.

Definition 3.4.2 *For every compact pair (X, A), let $\mathrm{Vect}(X, A)$ denote the set of isomorphism classes of elements of $\mathfrak{V}(X, A)$. The isomorphism class of (V, W, σ) in $\mathrm{Vect}(X, A)$ is denoted $[V, W, \sigma]$.*

Proposition 3.4.3 *Let (X, A) be a compact pair. Then $\mathrm{Vect}(X, A)$ is an abelian monoid.*

Proof Given elements $[V_1, W_1, \sigma_1]$ and $[V_2, W_2, \sigma_2]$ in $\mathrm{Vect}(X, A)$, define

$$[V_1, W_1, \sigma_1] + [V_2, W_2, \sigma_2] = [V_1 \oplus V_2, W_1 \oplus W_2, \sigma_1 \oplus \sigma_2].$$

Suppose that the triples (V_1, W_1, σ_1) and (V_2, W_2, σ_2) are isomorphic to $(\widetilde{V}_1, \widetilde{W}_1, \widetilde{\sigma}_1)$ and $(\widetilde{V}_2, \widetilde{W}_2, \widetilde{\sigma}_2)$ respectively. Then for $i = 1, 2$, there exist bundle isomorphisms $\nu_i : V_i \longrightarrow \widetilde{V}_i$ and $\omega_i : W_i \longrightarrow \widetilde{W}_i$ with the

property that the diagram

$$\begin{array}{ccc} V_i|A & \xrightarrow{\sigma_i} & W_i|A \\ \nu|A \downarrow & & \downarrow \omega|A \\ \widetilde{V}_i|A & \xrightarrow{\widetilde{\sigma}_i} & \widetilde{W}_i|A \end{array}$$

commutes. Then $\nu_1 \oplus \nu_2$ and $\omega_1 \oplus \omega_2$ are bundle isomorphisms that make the diagram

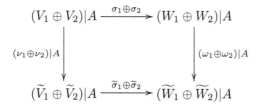

$$\begin{array}{ccc} (V_1 \oplus V_2)|A & \xrightarrow{\sigma_1 \oplus \sigma_2} & (W_1 \oplus W_2)|A \\ (\nu_1 \oplus \nu_2)|A \downarrow & & \downarrow (\omega_1 \oplus \omega_2)|A \\ (\widetilde{V}_1 \oplus \widetilde{V}_2)|A & \xrightarrow{\widetilde{\sigma}_1 \oplus \widetilde{\sigma}_2} & (\widetilde{W}_1 \oplus \widetilde{W}_2)|A \end{array}$$

commute, and therefore the addition of elements of $\mathrm{Vect}(X, A)$ does not depend upon the choice of representative for each isomorphism class. The class $[0, 0, \mathrm{id}]$ is the additive identity in $\mathrm{Vect}(X, A)$, and addition is commutative and associative because the internal Whitney sum of vector bundles has these properties up to isomorphism. $\qquad\square$

The reader might surmise that we next take the Grothendieck completion of $\mathrm{Vect}(X, A)$, but that does not give us the abelian group we want. Instead, we proceed in a different way.

Definition 3.4.4 *Let (X, A) be a compact pair. We declare an element of $\mathrm{Vect}(X, A)$ to be* trivial *if it can be written in the form $[V, V, \mathrm{id}]$, where V is a vector bundle over X and id is the identity map on $V|A$. The set of trivial elements of $\mathrm{Vect}(X, A)$ form a submonoid denoted $\mathrm{Triv}(X, A)$.*

Lemma 3.4.5 *Let (X, A) be a compact pair, let V be a vector bundle over X, and suppose $\sigma : V|A \longrightarrow V|A$ is homotopic through bundle isomorphisms to the identity map on $V|A$. Then there exists an element $[W, W, \mathrm{id}]$ of $\mathrm{Triv}(X, A)$ with the property that $[V, V, \sigma] + [W, W, \mathrm{id}]$ is trivial.*

Proof Imbed V into a trivial bundle $\Theta^n(X)$, choose a Hermitian metric on $\Theta^n(X)$, and let $V^\perp$ be the orthogonal complement bundle of V in

$\Theta^n(X)$. Then

$$[V, V, \sigma] + [V^\perp, V^\perp, \mathrm{id}] = [V \oplus V^\perp, V \oplus V^\perp, \sigma \oplus \mathrm{id}],$$

and $\sigma \oplus \mathrm{id}$ can be homotoped through bundle isomorphisms to the identity map on $(V \oplus V^\perp)|A$. Identify $V \oplus V^\perp$ with $\Theta^n(X)$ and let $\widetilde{\sigma}$ be the bundle isomorphism on $\Theta^n(A) = \Theta^n(X)|A$ that corresponds to $\sigma \oplus \mathrm{id}$. If we choose a basis for $\mathbb{C}^n$, then $\widetilde{\sigma}$ determines a matrix $\widetilde{\mathsf{S}}$ in $\mathrm{GL}(n, C(A))_0$. By Proposition 1.3.16 there exists a matrix $\widehat{\mathsf{S}}$ in $\mathrm{GL}(n, C(X))_0$ such that $\widehat{\mathsf{S}}|A = \widetilde{\mathsf{S}}$, and $\widehat{\mathsf{S}}$ determines a bundle isomorphism $\widehat{\sigma}$ from $\Theta^n(X)$ to itself. This gives us the commutative diagram

$$
\begin{array}{ccc}
\Theta^n(A) & \xrightarrow{\ \widetilde{\sigma}\ } & \Theta^n(A) \\
{\scriptstyle \widehat{\sigma}|A}\big\downarrow & & \big\downarrow{\scriptstyle \mathrm{id}} \\
\Theta^n(A) & \xrightarrow{\ \mathrm{id}\ } & \Theta^n(A),
\end{array}
$$

and therefore $[\Theta^n(X), \Theta^n(X), \widehat{\sigma}] = [V \oplus V^\perp, V \oplus V^\perp, \sigma \oplus \mathrm{id}]$ is in $\mathrm{Triv}(X, A)$. $\qquad\square$

Definition 3.4.6 *Let (X, A) be a compact pair. For any two elements $[V_1, W_1, \sigma_1]$ and $[V_2, W_2, \sigma_2]$ of $\mathrm{Vect}(X, A)$, define*

$$\mathrm{offdiag}(\sigma_1, \sigma_2) = \begin{pmatrix} 0 & \sigma_2 \\ \sigma_1 & 0 \end{pmatrix}.$$

Lemma 3.4.7 *Let (X, A) be a compact pair and suppose that $[V_1, W_1, \sigma_1]$ and $[V_2, W_2, \sigma_2]$ are elements of $\mathrm{Vect}(X, A)$. Then*

$$[V_1, W_1, \sigma_1] + [V_2, W_2, \sigma_2] = [V_1 \oplus V_2, W_2 \oplus W_1, \mathrm{offdiag}(\sigma_1, -\sigma_2)].$$

Proof From the definition of addition on $\mathrm{Vect}(X, A)$, we have

$$[V_1, W_1, \sigma_1] + [V_2, W_2, \sigma_2] = [V_1 \oplus V_2, W_1 \oplus W_2, \sigma_1 \oplus \sigma_2].$$

Define $\gamma : W_1 \oplus W_2 \longrightarrow W_2 \oplus W_1$ as $\gamma(w_1, w_2) = (-w_2, w_1)$. The map γ is a bundle isomorphism, and the diagram

$$
\begin{array}{ccc}
(V_1 \oplus V_2)|A & \xrightarrow{\ \sigma_1 \oplus \sigma_2\ } & (W_1 \oplus W_2)|A \\
{\scriptstyle \mathrm{id}}\big\downarrow & & \big\downarrow{\scriptstyle \gamma|A} \\
(V_1 \oplus V_2)|A & \xrightarrow{\hspace{2cm}} & (W_2 \oplus W_1)|A
\end{array}
$$

commutes, where offdiag$(\sigma_1, -\sigma_2)$ is the bottom arrow in the diagram. Therefore

$$(V_1 \oplus V_1, W_1 \oplus W_2, \sigma_1 \oplus \sigma_2) \cong (V_1 \oplus V_2, W_2 \oplus W_1, \text{offdiag}(\sigma_1, -\sigma_2)).$$

$\square$

Lemma 3.4.8 *Let (X, A) be a compact pair, let V and W be vector bundles over X, and suppose that $\sigma : V|A \longrightarrow W|A$ is a bundle isomorphism. Then the matrix* offdiag$(\sigma, -\sigma^{-1})$ *is homotopic through bundle isomorphisms on $(V \oplus W)|A$ to the identity map.*

Proof The product

$$\begin{pmatrix} \text{id} & -t\sigma^{-1} \\ 0 & \text{id} \end{pmatrix} \begin{pmatrix} \text{id} & 0 \\ t\sigma & \text{id} \end{pmatrix} \begin{pmatrix} \text{id} & -t\sigma^{-1} \\ 0 & \text{id} \end{pmatrix}$$

produces such a homotopy. $\square$

Proposition 3.4.9 *For every compact pair (X, A), the quotient monoid* $\text{Vect}(X, A)/\text{Triv}(X, A)$ *is an abelian group, and $[W, V, \sigma^{-1}]$ is the inverse of $[V, W, \sigma]$ for every element $[V, W, \sigma]$ of* $\text{Vect}(X, A)/\text{Triv}(X, A)$.

Proof Because elements of a monoid do not necessarily have inverses, we must be careful in defining the quotient monoid. In our case, elements $[V_1, W_1, \sigma_1]$ and $[V_2, W_2, \sigma_2]$ of $\text{Vect}(X, A)$ are in the same coset of $\text{Triv}(X, A)$ if and only if there exist vector bundles V' and W' over X such that

$$[V_1, W_1, \sigma_1] + [V', V', \text{id}] = [V_2, W_2, \sigma_2] + [W', W', \text{id}].$$

Take $[V, W, \sigma]$ in $\text{Vect}(X, A)$. Lemma 3.4.7 implies that

$$[V, W, \sigma] + [W, V, \sigma^{-1}] = [V \oplus W, V \oplus W, \text{offdiag}(\sigma, -\sigma^{-1})].$$

Lemmas 3.4.5 and 3.4.8 imply that we can add an element of $\text{Triv}(X, A)$ to $[V, W, \sigma] + [W, V, \sigma^{-1}]$ and obtain an element of $\text{Triv}(X, A)$, which yields the proposition. $\square$

Definition 3.4.10 *For each compact pair (X, A), set*

$$L(X, A) = \text{Vect}(X, A)/\text{Triv}(X, A).$$

We will abuse notation and let $[V, W, \sigma]$ denote both an element of $\text{Vect}(X, A)$ and the coset it defines in K$^0(X, A)$.

Proposition 3.4.11 *Let (X, A) be a compact pair, let $[V, W, \sigma_0]$ and $[V, W, \sigma_1]$ be elements of $\mathrm{L}(X, A)$, and suppose that σ_0 and σ_1 are homotopic through bundle isomorphisms from $V|A$ to $W|A$. Then $[V, W, \sigma_0] = [V, W, \sigma_1]$ in $\mathrm{L}(X, A)$.*

Proof From Lemma 3.4.7 and Proposition 3.4.9, we have

$$[V, W, \sigma_0] - [V, W, \sigma_1] = [V, W, \sigma_0] + [W, V, \sigma_1^{-1}]$$
$$= [V \oplus W, V \oplus W, \mathrm{offdiag}(-\sigma_1^{-1}, \sigma_0)].$$

Choose a homotopy $\{\sigma_t\}$ of bundle isomorphisms from σ_0 to σ_1. Then $\mathrm{offdiag}(-\sigma_1^{-1}, \sigma_t)$ is a homotopy of bundle isomorphisms. Concatenate this homotopy with the one from Lemma 3.4.8; the result then follows from Lemma 3.4.5 and the definition of $\mathrm{L}(X, A)$. $\qquad\square$

Proposition 3.4.12 *Let (X, A) be a compact pair. Then every element of $\mathrm{L}(X, A)$ can be written in the form $[V, \Theta^n(X), \sigma]$ for some vector bundle V and some natural number n.*

Proof Suppose X has components $X_1, X_2, \ldots, X_l$, and suppose for each $1 \leq k \leq l$, we have an element $[V_k, \Theta^{n_k}(X_k), \sigma_k]$ of $\mathrm{L}(X_k, A \cap X_k)$. Choose n to be larger than any of the n_ks. Then

$$[V_k, \Theta^{n_k}(X_k), \sigma_k] =$$
$$[V_k, \Theta^{n_k}(X_k), \sigma_k] + [\Theta^{n-n_k}(X_k), \Theta^{n-n_k}(X_k), \mathrm{id}\,|(A \cap X_k)]$$
$$= [V_k \oplus \Theta^{n-n_k}(X_k), \Theta^n(X_k)\sigma \oplus \mathrm{id}\,|(A \cap X_k)].$$

for each $1 \leq k \leq l$. Collectively these elements determine an element of $\mathrm{L}(X, A)$. Therefore we may assume for the rest of the proof that X is connected.

Take $[V, W, \sigma]$ in $\mathrm{L}(X, A)$. Imbed W into a trivial bundle $\Theta^n(X)$, equip $\Theta^n(X)$ with a Hermitian metric, and let $W^\perp$ be the orthogonal complement of W in $\Theta^n(X)$. Then

$$[V, W, \sigma] = [V, W, \sigma] + [W^\perp, W^\perp, \mathrm{id}]$$
$$= [V \oplus W^\perp, W \oplus W^\perp, \sigma \oplus \mathrm{id}],$$

and the desired result follows by identifying $W \oplus W^\perp$ with $\Theta^n(X)$. $\qquad\square$

Proposition 3.4.13 *The assignment $(X, A) \mapsto \mathrm{L}(X, A)$ is a contravariant functor from the category of compact pairs to the category of abelian groups.*

Proof Suppose that $\phi : (Y, B) \longrightarrow (X, A)$ is a morphism of compact pairs and let $[V, W, \sigma]$ be an element of L(X, A). Define $\phi^*\sigma :$ $(\phi^*V)|A \longrightarrow (\phi^*W)|A$ by the formula $(\phi^*\sigma)(a, v) = (\phi(a), \sigma(v))$. Then $[\phi^*V, \phi^*W, \phi^*\sigma]$ is an element of Vect(Y, B).

Suppose that (V, W, σ) is isomorphic to (V', W', σ'). Then there exist isomorphisms $\nu : V \longrightarrow V'$ and $\omega : W \longrightarrow W'$ that make the diagram

$$
\begin{array}{ccc}
(\phi^*V)|B & \xrightarrow{\phi^*\sigma} & (\phi^*W)|B \\
{\scriptstyle(\phi^*\nu)|B}\Big\downarrow & & \Big\downarrow{\scriptstyle(\phi^*\omega)|B} \\
(\phi^*V')|B & \xrightarrow{\phi^*\sigma'} & (\phi^*W')|B
\end{array}
$$

commute. Thus ϕ^* determines a well defined map from Vect(X, A) to Vect(Y, B). The fact that ϕ^* distributes across internal Whitney sum implies that ϕ^* is a monoid homomorphism and that ϕ^* maps Triv(X, A) to Triv(Y, B). Therefore $\phi^* :$ L$(X, A) \longrightarrow$ L(Y, B) is a group homomorphism, and the reader can easily verify that the axioms of a contravariant functor are satisfied. □

Lemma 3.4.14 *For every pointed space (X, x_0), there exists a natural isomorphism from* L(X, x_0) *to* $\widetilde{K}^0(X, x_0)$.

Proof Let j denote the inclusion of x_0 into X. We begin by constructing a homomorphism ψ that makes the sequence

$$
0 \longrightarrow \mathrm{L}(X, x_0) \xrightarrow{\ \psi\ } \mathrm{K}^0(X) \xrightarrow{\ j^*\ } \mathrm{K}^0(x_0)
$$

exact. Define ψ by the formula $\psi[V, W, \sigma] = [V] - [W]$. If (V_1, W_1, σ_1) is isomorphic to (V_2, W_2, σ_2), then there exist isomorphisms ν and ω from V_1 to V_2 and W_1 to W_2 respectively, and thus

$$
\psi[V_1, W_1, \sigma_1] = [V_1] - [W_1] = [V_2] - [W_2] = \psi[V_2, W_2, \sigma_1].
$$

We have $\psi[V, V, \mathrm{id}] = 0$ for every vector bundle V over X, and thus ψ is well defined. The definition of ψ immediately implies that it is a group homomorphism and that $j^*\psi = 0$. On the other hand, suppose that $j^*([V] - [W]) = 0$ in K$^0(x_0)$. The fibers of V and W over the point x_0 must have the same dimension, and therefore there exists a bundle isomorphism σ from $V|\{x_0\}$ to $W|\{x_0\}$. Thus $\psi[V, W, \sigma] = [V] - [W]$, and hence our sequence is exact at K$^0(X)$.

To show exactness at $L(X, x_0)$, suppose that

$$\psi[V, W, \sigma] = [V] - [W] = 0$$

in $K^0(X)$. Then there exists a vector bundle $\widetilde{V}$ and an isomorphism $\gamma : V \oplus \widetilde{V} \longrightarrow W \oplus \widetilde{V}$. By imbedding $V \oplus \widetilde{V}$ in a trivial bundle and replacing $\widetilde{V}$ by $\widetilde{V} \oplus (V \oplus \widetilde{V})^{\perp}$, we may assume that $V \oplus \widetilde{V}$ and $W \oplus \widetilde{V}$ are trivial bundles of some rank n. Fix isomorphisms $V \oplus \widetilde{V} \cong \Theta^n(X)$ and $W \oplus \widetilde{V} \cong \Theta^n(X)$; then γ determines an element in $\mathrm{GL}(n, C(X))$. We have the isomorphism

$$\sigma \oplus \mathrm{id} : (V \oplus \widetilde{V})|\{x_0\} \longrightarrow (W \oplus \widetilde{V})|\{x_0\},$$

and hence an isomorphism

$$\alpha_0 = (\gamma|\{x_0\})(\sigma \oplus \mathrm{id})^{-1} : (W \oplus \widetilde{V})|\{x_0\} \longrightarrow (W \oplus \widetilde{V})|\{x_0\}.$$

Restrict the isomorphism $W \oplus \widetilde{V} \cong \Theta^n(X)$ to $\{x_0\}$ and write α_0 as an element of $\mathrm{GL}(n, \mathbb{C})$. Proposition 1.3.11 states that α_0 is an element of $\mathrm{GL}(n, \mathbb{C})_0$, and Proposition 1.3.16 allows us to extend α_0 to an element of $\mathrm{GL}(n, C(X))$, which in turn gives us a bundle isomorphism α from $W \oplus \widetilde{V}$ to $W \oplus \widetilde{V}$. We have the commutative diagram

$$
\begin{array}{ccc}
(V \oplus \widetilde{V})|\{x_0\} & \xrightarrow{\;\sigma \oplus \mathrm{id}\;} & (W \oplus \widetilde{V})|\{x_0\} \\
{\scriptstyle \gamma|\{x_0\}}\Big\downarrow & & \Big\downarrow{\scriptstyle \alpha|\{x_0\}} \\
(W \oplus \widetilde{V})|\{x_0\} & \xrightarrow{\;\mathrm{id}\;} & (W \oplus \widetilde{V})|\{x_0\},
\end{array}
$$

and therefore

$$[V, W, \sigma] = [V, W, \sigma] + [\widetilde{V}, \widetilde{V}, \mathrm{id}] =$$
$$[V \oplus \widetilde{V}, W \oplus \widetilde{V}, \sigma \oplus \mathrm{id}] = [W \oplus \widetilde{V}, W \oplus \widetilde{V}\,\mathrm{id}] = 0$$

in $L(X, x_0)$.

We also have a commutative diagram

$$
\begin{array}{ccccccc}
0 & \longrightarrow & L(X, x_0) & \xrightarrow{\;\psi\;} & K^0(X) & \xrightarrow{\;j^*\;} & K^0(x_0) \\
& & {\scriptstyle =}\Big\| & & \Big\downarrow & & \Big\downarrow{\scriptstyle =} \\
0 & \longrightarrow & K^0(X, x_0) & \xrightarrow{\;\mu_1^*\;} & K^0(X) & \xrightarrow{\;j^*\;} & K^0(x_0)
\end{array}
$$

with exact rows, and Proposition 1.8.9 implies that the assignment

$[V, W, \sigma] \mapsto [V] - [W]$ is well defined and is an isomorphism from $\mathrm{L}(X, x_0)$ to $\mathrm{K}^0(X, x_0)$. Finally, this isomorphism is natural because the other maps in the diagram are natural. $\qquad\square$

Theorem 3.4.15 *There exists a natural isomorphism* $\chi : \mathrm{L}(X, A) \longrightarrow \mathrm{K}^0(X, A)$ *for every compact pair* (X, A).

Proof Let $[V, W, \sigma]$ be an element of $\mathrm{L}(X, A)$ and form the clutched bundle $V \cup_\sigma W$ over $\mathcal{D}(X, A)$. Let $\psi : \mathcal{D}(X, A) \longrightarrow X$ be the map that identifies the two copies of X in $\mathcal{D}(X, A)$. Then the short exact sequence

$$0 \longrightarrow \mathrm{K}^0(X, A) \xrightarrow{\ \iota\ } \mathrm{K}^0(\mathcal{D}(X, A)) \overset{\mu_1^*}{\underset{\psi^*}{\rightleftarrows}} \mathrm{K}^0(X) \longrightarrow 0$$

that defines $\mathrm{K}^0(X, A)$ is split exact, and we use Proposition 1.8.7 to define

$$\chi[V, W, \sigma] = [V \cup_\sigma W] - \psi^* \mu_1^* [V \cup_\sigma W]$$

for every element $[V, W, \sigma]$ of $\mathrm{L}(X, A)$.

Suppose that $(V, W, \sigma) \cong (\widetilde{V}, \widetilde{W}, \widetilde{\sigma})$ and choose isomorphisms $\nu : V \longrightarrow \widetilde{V}$ and $\omega : W \longrightarrow \widetilde{W}$ so that the diagram

$$
\begin{array}{ccc}
V|A & \xrightarrow{\ \sigma\ } & W|A \\
{\scriptstyle \nu|A}\big\downarrow & & \big\downarrow{\scriptstyle \omega|A} \\
\widetilde{V}|A & \xrightarrow{\ \widetilde{\sigma}\ } & \widetilde{W}|A
\end{array}
$$

commutes. Then ν and ω define a bundle ismorphism

$$\nu \cup_\sigma \omega : V \cup_\sigma W \longrightarrow \widetilde{V} \cup_{\widetilde{\sigma}} \widetilde{W}$$

with inverse $\nu^{-1} \cup_\sigma \omega^{-1}$. Thus $[V \cup_\sigma W] = [\widetilde{V} \cup_{\widetilde{\sigma}} \widetilde{W}]$ in $\mathrm{K}^0(\mathcal{D}(X, A))$, and so χ does not depend on the isomorphism class of (V, W, σ). Second, for each vector bundle V over X we see that $\psi^* V = V \cup_{\mathrm{id}} V$, whence $\chi[V, V, \mathrm{id}] = 0$. Therefore χ is well defined. Furthermore, let $[V, W, \sigma]$ and $[\widetilde{V}, \widetilde{W}, \widetilde{\sigma}]$ be arbitrary elements of $\mathrm{L}(X, A)$. Then

$$(V \oplus \widetilde{V}) \cup_{\sigma \oplus \widetilde{\sigma}} (W \oplus \widetilde{W}) \cong (V \cup_\sigma \widetilde{V}) \oplus (W \cup_\sigma \widetilde{W}),$$

which implies that χ is a homomorphism.

Suppose that $\phi : (Y, B) \longrightarrow (X, A)$ is a morphism of compact pairs and take $[V, W, \sigma]$ in $\mathrm{L}(X, A)$. From the definitions of pullback and clutching we find that $\phi^*(V \cup_\sigma W)$ is isomorphic to $(\phi^*V) \cup_{\sigma\phi} (\phi^*W)$, which implies that the diagram

$$
\begin{array}{ccc}
L(X, A) & \xrightarrow{\ \chi\ } & K^0(X, A) \\
\phi^* \downarrow & & \downarrow \phi^* \\
L(Y, B) & \xrightarrow{\ \chi\ } & K^0(Y, B)
\end{array}
$$

commutes, and therefore χ is a natural transformation.

Let $q : (X, A) \longrightarrow ((X \backslash A)^+, \infty)$ be the quotient morphism. Using Proposition 2.5.3 to identify the reduced group $\widetilde{K}^0((X \backslash A)^+, \infty)$ with the relative group $K^0((X \backslash A)^+, \infty)$, we obtain a commutative diagram

$$
\begin{array}{ccc}
L((X \backslash A)^+, \infty) & \xrightarrow{\ \chi\ } & K^0((X \backslash A)^+, \infty) \\
q^* \downarrow & & \downarrow q^* \\
L(X, A) & \xrightarrow{\ \chi\ } & K^0(X, A).
\end{array}
$$

The homomorphisms on the top and right of the diagram are isomorphisms, and therefore χq^* is also an isomorphism, whence χ is surjective. To show that χ is injective, suppose that $\chi[V, W, \sigma] = 0$ in $K^0(X, A)$. Then $[V \cup_\sigma W] = \psi^* \mu_1^*[V \cup_\sigma W]$ in $K^0(\mathcal{D}(X, A))$, and therefore there exists a vector bundle $\widehat{W}$ over $\mathcal{D}(X, A)$ such that

$$
(V \cup_\sigma W) \oplus \widehat{W} \cong (V \cup_{\mathrm{id}} V) \oplus \widehat{W}.
$$

Distributing $\widehat{W}$, we have a bundle isomorphism

$$
\gamma : \left(V \oplus \widehat{W}\right) \cup_{\sigma \oplus \mathrm{id}} \left(W \oplus \widehat{W}\right) \longrightarrow \left(V \oplus \widehat{W}\right) \cup_{\mathrm{id} \oplus \mathrm{id}} \left(V \oplus \widehat{W}\right).
$$

Let γ_1 and γ_2 be the restrictions of γ to the two copies of X in $\mathcal{D}(X, A)$. The compatibility of γ with the clutching maps implies that the diagram

$$
\begin{array}{ccc}
(V \oplus \widehat{W})|A & \xrightarrow{\ \sigma \oplus \mathrm{id}\ } & (W \oplus \widehat{W})|A \\
\gamma_1|A \downarrow & & \downarrow \gamma_2|A \\
(V \oplus \widehat{W})|A & \xrightarrow{\ \sigma \oplus \mathrm{id}\ } & (V \oplus \widehat{W})|A
\end{array}
$$

commutes. Therefore

$$[V \oplus \widehat{W}, W \oplus \widehat{W}, \sigma \oplus \mathrm{id}] = [V \oplus \widehat{W}, V \oplus \widehat{W}, \mathrm{id} \oplus \mathrm{id}]$$
$$[V, W, \sigma] + [\widehat{W}, \widehat{W}, \mathrm{id}] = [V, V, \sigma] + [\widehat{W}, \widehat{W}, \mathrm{id}]$$
$$[V, W, \sigma] = [V, V, \mathrm{id}],$$

and so $[V, W, \sigma] = 0$ in $\mathrm{K}^0(X, A)$, as desired. $\qquad\qquad\square$

In light of Theorem 3.4.15, we will use the notation $\mathrm{K}^0(X, A)$ for both our original $\mathrm{K}^0(X, A)$ and also for $\mathrm{L}(X, A)$; the reader should be able to tell from the context which picture of relative K-theory we are using.

We record a special case of Theorem 3.4.15 that we will need later in the chapter.

Proposition 3.4.16 *Let* $\widetilde{\mathbf{b}}$ *be the element of* $\mathrm{K}^0(B^2, S^1)$ *constructed in Proposition 3.3.9. Then* $\chi[\Theta^1(B^2), \Theta^1(B^2), z] = \widetilde{\mathbf{b}}$.

Proof We maintain the notation of Theorem 3.4.15. Every vector bundle over B^2 is trivial, which implies that

$$\chi[\Theta^1(B^2), \Theta^1(B^2), z] = [\Theta^1(B^2) \cup_z \Theta^1(B^2)] - \psi^* \mu_1^* [\Theta^1(B^2) \cup_z \Theta^1(B^2)]$$
$$= [\Theta^1(B^2) \cup_z \Theta^1(B^2)] - [1, 1].$$

Take one copy of B^2 to be the closed unit disk in $\mathbb{C}$ and take a second copy of B^2 to be (homeomorphic to) the complement of the open unit disk in the one-point compactification $\mathbb{C}^+$ of $\mathbb{C}$. Then $\Theta^1(B^2) \cup_z \Theta^1(B^2)$ is a vector bundle over $\mathbb{C}^+ \cong \mathcal{D}(B^2, S^1)$. Extend

$$\begin{pmatrix} |z|^2 & z\sqrt{1 - |z|^2} \\ \bar{z}\sqrt{1 - |z|^2} & 1 - |z|^2 \end{pmatrix}$$

to an idempotent E over all of $\mathbb{C}^+$ by setting it equal to $\mathrm{diag}(1, 0)$ outside of the unit disk. Inside the unit disk, the range of E at a point z is the span of the vector $(z, \sqrt{1 - |z|^2})$, and the definition of clutching then implies that

$$\mathrm{Ran}\,\mathsf{E} \cong \Theta^1(B^2) \cup_z \Theta^1(B^2),$$

which yields the desired result. $\qquad\qquad\square$

Theorem 3.4.17 *Let* (X, A) *and* (Y, B) *be compact pairs and suppose that* $[V_0, V_1, \alpha]$ *and* $[W_0, W_1, \beta]$ *are elements of* $\mathrm{L}(X, A)$ *and* $\mathrm{L}(Y, B)$ *respectively. Choose bundle homomorphisms* $\widetilde{\alpha}$ *and* $\widetilde{\beta}$ *so that* $\widetilde{\alpha}|A = \alpha$

and $\widetilde{\beta}|B = \beta$, equip the vector bundles V_0, V_1, W_0, and W_1 with Hermitian metrics, and let $\widetilde{\alpha}^* : V_1 \longrightarrow V_0$ and $\widetilde{\beta}^* : W_1 \longrightarrow W_0$ be the adjoint maps. Define

$$\gamma = \begin{pmatrix} \widetilde{\alpha} \boxtimes I & -I \boxtimes \widetilde{\beta}^* \\ I \boxtimes \widetilde{\beta} & \widetilde{\alpha}^* \boxtimes I \end{pmatrix}.$$

Then

$$[V_0, V_1, \alpha] \boxtimes [W_0, W_1, \beta] =$$
$$\left[(V_0 \boxtimes W_0) \oplus (V_1 \boxtimes W_1), (V_1 \boxtimes W_0) \oplus (V_0 \boxtimes W_1), \gamma \right]$$

is a formula for the external product

$$L(X, A) \otimes L(Y, B) \longrightarrow L\big(X \times Y, (X \times B) \cup (A \times Y)\big).$$

Proof Proposition 1.5.19 guarantees the existence of $\widetilde{\alpha}$ and $\widetilde{\beta}$. To show that γ is an isomorphism when restricted to $(X \times B) \cup (A \times Y)$, first note that if (x, y) is a point in $X \times Y$ and that $\gamma_{(x,y)}\gamma^*_{(x,y)}$ is an isomorphism from $(V_1)_x \boxtimes (W_0)_y$ to itself, then $\gamma_{(x,y)}$ is surjective. Moreover, because $(V_1)_x \boxtimes (W_0)_y$ is finite-dimensional, the vector space homomorphism $\gamma_{(x,y)}$ is an isomorphism. Because

$$\gamma_{(x,y)}\gamma^*_{(x,y)} = \begin{pmatrix} \widetilde{\alpha}_x\widetilde{\alpha}^*_x \boxtimes I + I \boxtimes \widetilde{\beta}^*_y\widetilde{\beta}_y & 0 \\ 0 & \widetilde{\alpha}^*_x\widetilde{\alpha}_x \boxtimes I + I \boxtimes \widetilde{\beta}_y\widetilde{\beta}^*_y \end{pmatrix},$$

we see that $\gamma_{(x,y)}$ is an isomorphism precisely when $\widetilde{\alpha}_x\widetilde{\alpha}^*_x \boxtimes I + I \boxtimes \widetilde{\beta}^*_y\widetilde{\beta}_y$ and $\widetilde{\alpha}^*_x\widetilde{\alpha}_x \boxtimes I + I \boxtimes \widetilde{\beta}_y\widetilde{\beta}^*_y$ are both isomorphisms.

Suppose that $(\widetilde{\alpha}_x\widetilde{\alpha}^*_x \boxtimes I + I \boxtimes \widetilde{\beta}^*_y\widetilde{\beta})\omega = 0$ for some ω in $(V_1)_x \boxtimes (W_0)_y$. Then

$$0 = \langle (\widetilde{\alpha}_x\widetilde{\alpha}^*_x \boxtimes I + I \boxtimes \widetilde{\beta}^*_y\widetilde{\beta})\omega, \omega \rangle$$
$$= \langle (\widetilde{\alpha}_x\widetilde{\alpha}^*_x \boxtimes I)\omega, \omega \rangle + \langle (I \boxtimes \widetilde{\beta}^*_y\widetilde{\beta})\omega, \omega \rangle$$
$$= \langle (\widetilde{\alpha}^*_x \boxtimes I)\omega, (\widetilde{\alpha}^*_x \boxtimes I)\omega \rangle + \langle (I \boxtimes \widetilde{\beta}_y)\omega, (I \boxtimes \widetilde{\beta}_y)\omega \rangle.$$

Both summands are nonnegative and Definition 1.1.1 implies that both $(\widetilde{\alpha}^*_x \boxtimes I)\omega$ and $(I \boxtimes \widetilde{\beta}_y)\omega$ are zero. The homomorphism $I \boxtimes \widetilde{\beta}_y$ is an isomorphism when y is in B, and Proposition 1.1.9 implies that $(\widetilde{\alpha}^*_x \boxtimes I)$ is an isomorphism when x is in A. Therefore $\widetilde{\alpha}_x\widetilde{\alpha}^*_x \boxtimes I + I \boxtimes \widetilde{\beta}^*_y\widetilde{\beta}_y$ is an isomorphism when restricted to $(X \times B) \cup (A \times Y)$, and a similar argument shows that the same is true of $\widetilde{\alpha}^*_x\widetilde{\alpha}_x \boxtimes I + I \boxtimes \widetilde{\beta}_y\widetilde{\beta}^*_y$. Hence

$$\left[(V_0 \boxtimes W_0) \oplus (V_1 \boxtimes W_1), (V_1 \boxtimes W_0) \oplus (V_0 \boxtimes W_1), \gamma \right]$$

is an element of $L\big(X \times Y, (X \times B) \cup (A \times Y)\big)$, and the uniquess of the external product completes the proof. $\square$

3.5 The exterior algebra

To prove many of the main results of this chaper, we need a vector bundle construction called the *exterior product*. As with the tensor product, we begin with vector spaces.

Definition 3.5.1 *Let $\mathcal{V}$ be a vector space, define $\bigotimes^0 \mathcal{V} = \mathbb{C}$, and for each positive integer k, define $\bigotimes^k \mathcal{V}$ to be the tensor product of k copies of $\mathcal{V}$. The* tensor algebra *of $\mathcal{V}$ is the set*

$$\mathcal{T}(\mathcal{V}) = \bigoplus_{k=0}^{\infty} \left(\bigotimes{}^k \mathcal{V} \right).$$

The tensor algebra is an algebra with unit; the scalar multiplication and addition come from the corresponding operations on each $\bigotimes^k \mathcal{V}$, tensor product serves as the multiplication, and 1 is the multiplicative identity.

We are primarily interested not in $\mathcal{T}(\mathcal{V})$, but a quotient algebra of it.

Definition 3.5.2 *Let $\mathcal{V}$ be a vector space and let $\mathcal{J}(\mathcal{V})$ be the ideal in $\mathcal{T}(\mathcal{V})$ generated by all simple tensors of the form $v \otimes v$; i.e.,*

$$\mathcal{J}(\mathcal{V}) = \{\tau \otimes v \otimes v \otimes \mu : \tau, \mu \in \mathcal{T}(\mathcal{V}), v \in \mathcal{V}\}.$$

The exterior algebra *of $\mathcal{V}$ is the quotient algebra*

$$\bigwedge(\mathcal{V}) = \mathcal{T}(\mathcal{V})/\mathcal{J}(\mathcal{V}).$$

The binary operation $\wedge$ in $\bigwedge(\mathcal{V})$ is called the exterior product *or* wedge product. *For each nonnegative integer k, we let $\bigwedge^k(\mathcal{V})$ be the image of $\bigotimes^k \mathcal{V}$ in $\bigwedge(\mathcal{V})$ and say that elements of $\bigwedge^k(\mathcal{V})$ have* degree k. *We also set*

$$\bigwedge{}^{even}(\mathcal{V}) = \bigoplus_{k \ even} \bigwedge{}^k(\mathcal{V})$$

$$\bigwedge{}^{odd}(\mathcal{V}) = \bigoplus_{k \ odd} \bigwedge{}^k(\mathcal{V}).$$

Finally, for each simple tensor $v_1 \otimes v_2 \otimes \cdots \otimes v_k$, we denote its image in $\bigwedge(\mathcal{V})$ by $v_1 \wedge v_2 \wedge \cdots \wedge v_k$ and call such an element a simple wedge.

Proposition 3.5.3 *Let V be a vector space. For all complex numbers λ and elements ω, $\tilde{\omega}$, ζ and $\tilde{\zeta}$ in $\bigwedge(V)$:*

(i) $(\omega + \tilde{\omega}) \wedge \zeta = \omega \wedge \zeta + \tilde{\omega} \wedge \zeta;$

(ii) $\omega \wedge (\zeta + \tilde{\zeta}) = \omega \wedge \zeta + \omega \wedge \tilde{\zeta};$

(iii) $(\lambda \omega) \wedge \zeta = \omega \wedge (\lambda \zeta) = \lambda(\omega \wedge \zeta).$

In addition, simple wedges $v_1 \wedge v_2 \wedge \cdots \wedge v_k$ equal zero if $v_1, v_2, \ldots, v_k$ are not distinct, or, more generally, when $v_1, v_2, \ldots, v_k$ are linearly dependent.

Proof Apply Definitions 3.2.5 and 3.5.2. $\qquad\square$

Note that Proposition 3.5.3 implies that $\bigwedge^n(V) = 0$ whenever n is greater than the rank of V.

Proposition 3.5.4 *Let V be a vector space, let k and l be nonnegative integers, and suppose that ω and ζ are in $\bigwedge^k(V)$ and $\bigwedge^l(V)$ respectively. Then $\zeta \wedge \omega = (-1)^{kl} \omega \wedge \zeta$.*

Proof Using Proposition 3.5.3 and induction, we need only show that $w \wedge v = -v \wedge w$ for all v and w in V, which follows from the equations

$$
\begin{aligned}
0 &= (v + w) \wedge (v + w) \\
&= v \wedge v + v \wedge w + w \wedge v + w \wedge w \\
&= v \wedge w + w \wedge v.
\end{aligned}
$$

$\qquad\square$

Proposition 3.5.5 *Suppose that $\{e_1, e_2, \ldots, e_n\}$ is a basis for a vector space V. Then for each $1 < k \leq n$, the set*

$$\{e_{i_1} \wedge e_{i_2} \wedge \cdots \wedge e_{i_k} : i_1 < i_2 < \cdots < i_k\}$$

is a basis for the vector space $\bigwedge^k(V)$, and therefore has dimension $\binom{n}{k}$.

Proof Write elements of V in terms of the basis, and observe that

$$\sum_{i=1}^n \alpha_i e_i \otimes \sum_{i=1}^n \alpha_i e_i = \sum_{i=1}^n \alpha_i^2 e_i \otimes e_i + \sum_{i<j} \alpha_i \alpha_j (e_i \otimes e_j + e_j \otimes e_i).$$

Elements of $\mathcal{J}(V)$ are linear combinations of quantities of the form that appear on the right-hand side of this equation, which implies that if

$\sum_{i,j=1}^{n} \beta_{i,j} e_i \otimes e_j$ is in $\mathcal{J}(\mathcal{V})$, we must have $\beta_{i,j} = \beta_{j,i}$ for all i and j. In particular, $e_i \otimes e_j$ is not in $\mathcal{J}(\mathcal{V})$ when $i \neq j$.

For $k > 2$, let S_k denote the symmetric group. For each σ in S_k, let $\text{sign}(\sigma)$ denotes its sign; i.e., we take $\text{sign}(\sigma)$ to be 1 if σ is an odd permutation and 0 if σ is even. Then a computation shows that if

$$\sum_{1 \leq i_1, i_2, \ldots, i_k \leq n} \beta_{i_1, i_2, \ldots, i_k} e_{i_1} \otimes e_{i_2} \otimes \cdots \otimes e_{i_k}$$

is in $\mathcal{J}(\mathcal{V})$, then

$$\sum_{\sigma \in S_k} (-1)^{\text{sign}(\sigma)} \beta_{i_{\sigma(1)}, i_{\sigma(2)}, \ldots, i_{\sigma(k)}} = 0$$

for all k-tuples $(i_1, i_2, \ldots, i_k)$. Thus $e_{i_1} \otimes e_{i_2} \otimes \cdots \otimes e_{i_k}$ is not in $\mathcal{J}(\mathcal{V})$ whenever $i_1, i_2, \ldots, i_k$ are distinct, and therefore the elements of the exterior algebra that appear in the statement of the proposition are all nonzero. Clearly such elements span $\bigwedge^k(\mathcal{V})$; to show that they are linearly independent, suppose that

$$\sum_{i_1 < i_2 < \cdots < i_k} \alpha_{i_1, i_2, \ldots, i_k} e_{i_1} \wedge e_{i_2} \wedge \cdots \wedge e_{i_k} = 0.$$

Fix a k-tuple $(i_1, i_2, \ldots, i_k)$ of distinct elements and let $e_{i_{k+1}}, \ldots, e_{i_n}$ be the remaining basis elements of $\mathcal{V}$. By taking the exterior product with $e_{i_{k+1}} \wedge \cdots \wedge e_{i_n}$ on both sides of the above equation, we see that all the terms except the one we have singled out vanish, and Proposition 3.5.4 gives us

$$\pm \alpha_{i_1, i_2, \ldots, i_k} e_1 \wedge e_2 \wedge \cdots \wedge e_n = 0,$$

whence $\alpha_{i_1, i_2, \ldots, i_k} = 0$. Our choice of k-tuple with distinct elements was arbitrary, and thus we have linear independence. Finally, we see that the dimension for $\bigwedge^k(\mathcal{V})$ equals the number of ways to choose k objects out of n with a prescribed order; i.e., the number of combinations of n objects taken k at a time. $\square$

The next result shows that the exterior algebra behaves nicely with respect to direct sums.

Proposition 3.5.6 *Let $\mathcal{V}$ and $\mathcal{W}$ be vector spaces. For each natural number n, there exists an isomorphism*

$$\bigwedge^n(\mathcal{V} \oplus \mathcal{W}) \cong \bigoplus_{k+l=n} \bigwedge^k(\mathcal{V}) \otimes \bigwedge^l(\mathcal{W}).$$

Proof Let $\{e_1, e_2, \ldots, e_m\}$ and $\{f_1, f_2, \ldots, f_n\}$ be bases for $\mathcal{V}$ and $\mathcal{W}$ respectively. Then the set

$$\{(e_1, 0), (e_2, 0), \ldots, (e_m, 0), (0, f_1), \ldots, (0, f_m)\}$$

is a basis for $\mathcal{V} \oplus \mathcal{W}$, and we define

$$\tau\big((e_{i_1}, e_{i_2}, \ldots, e_{i_k}) \otimes (f_{j_1}, f_{j_2}, \ldots, f_{j_l})\big) =$$
$$(e_{i_1}, 0) \wedge (e_{i_2}, 0) \wedge \cdots \wedge (e_{i_k}, 0) \wedge (0, f_{j_1}) \wedge (0, f_{j_2}) \wedge \cdots \wedge (0, f_{j_l})$$

for all k-tuples $(e_{i_1}, e_{i_2}, \ldots, e_{i_k})$ and l-tuples $(f_{j_1}, f_{j_2}, \ldots, f_{j_l})$ with $k + l = n$. We can extend τ by linearity, and Example 3.2.8 implies that τ is an isomorphism. Furthermore, by writing everything out in terms of the bases, we find that

$$\tau\big((v_1, v_2, \ldots, v_k) \otimes (w_1, w_2, \ldots, w_l)\big) =$$
$$(v_1, 0) \wedge (v_2, 0) \wedge \cdots \wedge (v_k, 0) \wedge (0, w_1) \wedge (0, w_2) \wedge \cdots \wedge (0, w_l)$$

for all k-tuples $(v_1, v_2, \ldots, v_k)$ and l-tuples $(w_1, w_2, \ldots, w_l)$ of elements for $\mathcal{V}$ and $\mathcal{W}$ respectively, which shows that our isomorphism does not depend on the choice of bases. $\qquad\square$

We next put an inner product on the exterior algebra.

Proposition 3.5.7 *Let $\mathcal{V}$ be an inner product space. There exists a unique inner product on $\bigwedge(\mathcal{V})$ with the following property: given simple tensors $v_1 \wedge v_2 \wedge \cdots \wedge v_k$ and $w_1 \wedge w_2 \wedge \cdots \wedge w_l$ in $\bigwedge(\mathcal{V})$, let $\{\langle v_i, w_j \rangle\}$ denote the k by l matrix whose (i, j) entry is $\langle v_i, w_j \rangle$. Then*

$$\langle v_1 \wedge v_2 \wedge \cdots \wedge v_k, w_1 \wedge w_2 \wedge \cdots \wedge w_l \rangle = \begin{cases} 0 & \text{if } k \neq l \\ \det\{\langle v_i, w_j \rangle\} & \text{if } k = l. \end{cases}$$

Proof Take $\{1\}$ to be an orthonormal basis for $\mathbb{C} = \bigwedge^0(\mathcal{V})$, fix an orthonormal basis $\{e_1, e_2, \ldots, e_m\}$ of $\mathcal{V} = \bigwedge^1(\mathcal{V})$, and declare the basis defined in Proposition 3.5.5 to be orthonormal for each $1 < k \leq m$. We then obtain an inner product on all of $\bigwedge(\mathcal{V})$ by decreeing that the vector subspaces $\bigwedge^k(\mathcal{V})$ are pairwise orthogonal.

The inner product of any two of the orthonormal basis elements of $\bigwedge(\mathcal{V})$ satisfies the desired formula. To show the formula holds in general, write each element of $\bigwedge(\mathcal{V})$ as a linear combination of the orthonormal basis, use Propositions 3.5.3 and 3.5.4 to decompose the inner products of arbitrary elements of $\bigwedge(\mathcal{V})$ into linear combinations of inner products of the orthonormal basis elements, and repeatedly invoke the following two properties of the determinant:

(i) the interchange of any two rows of a matrix changes the determinant by a factor of -1;

(ii)

$$
\det \begin{pmatrix}
\lambda a_{11} + \lambda' a'_{11} & \lambda a_{12} + \lambda' a'_{12} & \cdots & \lambda a_{1n} + \lambda' a'_{1n} \\
a_{21} & a_{22} & \cdots & a_{2n} \\
\vdots & \vdots & \ddots & \vdots \\
a_{n1} & a_{n2} & \cdots & a_{nn}
\end{pmatrix} =
$$

$$
\lambda \det \begin{pmatrix}
a_{11} & a_{12} & \cdots & a_{1n} \\
a_{21} & a_{22} & \cdots & a_{2n} \\
\vdots & \vdots & \ddots & \vdots \\
a_{n1} & a_{n2} & \cdots & a_{nn}
\end{pmatrix} + \lambda' \det \begin{pmatrix}
a'_{11} & a'_{12} & \cdots & a'_{1n} \\
a_{21} & a_{22} & \cdots & a_{2n} \\
\vdots & \vdots & \ddots & \vdots \\
a_{n1} & a_{n2} & \cdots & a_{nn}
\end{pmatrix}.
$$

The inner product we have constructed is unique because we can write every element of $\bigwedge(\mathcal{V})$ as a unique linear combination of our orthonormal basis. $\qquad\square$

We next extend the idea of exterior product to vector bundles.

Definition 3.5.8 *Suppose that $\mathcal{V}$ and $\mathcal{W}$ are vector spaces and that $\phi : \mathcal{V} \longrightarrow \mathcal{W}$ is a linear transformation. Then ϕ extends to a linear map on $\bigwedge(\mathcal{V})$ by declaring*

$$
\phi(v_1 \wedge v_2 \wedge \cdots \wedge v_k) = \phi(v_1) \wedge \phi(v_2) \wedge \cdots \wedge \phi(v_k)
$$

for all simple tensors and extending ϕ linearly to all of $\bigwedge(\mathcal{V})$.

Construction 3.5.9 *Let V be a vector bundle over a compact Hausdorff space X, use Proposition 1.7.9 to imbed V into a trivial bundle $\Theta^n(X)$, and apply Corollary 1.7.13 to write $V = \operatorname{Ran} \mathsf{E}$ and for some idempotent $\mathsf{E} : \Theta^n(X) \longrightarrow V$ in $\mathrm{M}(n, C(X))$. For each integer k, apply the rule in Definition 3.5.8 to obtain an idempotent*

$$
\bigwedge\nolimits^k(\mathsf{E}) : \bigwedge\nolimits^k(\Theta^n(X)) \longrightarrow \bigwedge\nolimits^k(V);
$$

observe that

$$
\bigwedge\nolimits^k(\mathsf{SES}^{-1}) = \bigwedge\nolimits^k(\mathsf{S}) \bigwedge\nolimits^k(\mathsf{E}) \bigwedge\nolimits^k(\mathsf{S}^{-1}) = \bigwedge\nolimits^k(\mathsf{S}) \bigwedge\nolimits^k(\mathsf{E}) \left(\bigwedge\nolimits^k(\mathsf{S})\right)^{-1}
$$

for every invertible S. Take the standard basis $\{e_1, e_2, \ldots, e_n, e_{n+1}\}$ of $\mathbb{C}^{n+1}$, use Proposition 3.5.5 to obtain a basis for $\bigwedge^k(\mathbb{C}^{n+1})$, and view *these basis elements as elements of $\bigwedge^k(\Theta^{n+1}(X))$ that are constant in X. The idempotent*

$$\bigwedge^k(\mathrm{diag}(\mathsf{E},0)) : \bigwedge^k(\Theta^{n+1}(X)) \longrightarrow \bigwedge^k(V)$$

kills basis elements of $\bigwedge^k(\Theta^{n+1}(X))$ that end in e_{n+1}, and therefore Lemma 1.7.1 gives us an invertible matrix U with complex entries such that

$$\bigwedge^k(\mathrm{diag}(\mathsf{E},0)) = \mathsf{U}\,\mathrm{diag}(\bigwedge^k(\mathsf{E}),0)\mathsf{U}^{-1}.$$

We define $\bigwedge^k(V)$ to be the vector bundle $\mathrm{Ran}\,\bigwedge^k(\mathsf{E})$; Proposition 1.7.6 implies that $\bigwedge^k(V)$ is well defined up to isomorphism. We then set

$$\bigwedge(V) = \bigoplus_{k=0}^{\infty}\bigwedge^k(V),$$

which makes sense because only a finite number of the vector bundles $\bigwedge^k(V)$ are nonzero. We define $\bigwedge^{even}(V)$ and $\bigwedge^{odd}(V)$ in a similar fashion. Finally, if V is a Hermitian vector bundle, we make $\bigwedge(V)$ into a Hermitian vector bundle by applying the inner product in Proposition 3.5.7 to each fiber.

We use the exterior product to define an object we will need in subsequent sections.

Definition 3.5.10 *Let X be compact Hausdorff and let $\mathrm{K}^0(X)[[t]]$ denote the ring of power series in t with elements of $\mathrm{K}^0(X)$ as coefficients. For each vector bundle V over X, define*

$$\bigwedge_t(V) = \sum_{k=0}^{\infty}[\bigwedge^k(V)]t^k.$$

The power series $\bigwedge_t(V)$ is in fact a polynomial because $\bigwedge^k(V)$ is trivial for k sufficiently large. As the next propostion shows, these power series are particularly well behaved with respect to internal Whitney sum.

Proposition 3.5.11 *Let V and W be vector spaces over a compact Hausdorff space. Then $\bigwedge_t(V \oplus W) = \bigwedge_t(V) \otimes \bigwedge_t(W)$.*

Proof Proposition 3.5.6 and Construction 3.5.9 imply that

$$\bigwedge^n(V \oplus W) \cong \bigoplus_{k+l=n}\bigwedge^k(V) \otimes \bigwedge^l(W)$$

for every nonnegative integer n; the proposition follows by multipying out $\bigwedge_t(V) \otimes \bigwedge_t(W)$. $\qquad\qquad\qquad\qquad\qquad\qquad\qquad\square$

3.6 Thom isomorphism theorem

Suppose that V is a vector bundle over a compact Hausdorff space X. Then V is a locally compact Hausdorff space, and we might hope that the K-theory of V is somehow related to the K-theory of X. For example, if $V = \Theta^n(X)$ for some natural number n, then V is homeomorphic to $X \times \mathbb{R}^{2n}$, and the Bott periodicity theorem and Theorem 2.6.13 imply that there are natural group isomorphisms $\mathrm{K}^p(V) \cong \mathrm{K}^p(X)$ for $p = 0, -1$. In fact, as we shall show in this chapter, the groups $\mathrm{K}^p(V)$ and $\mathrm{K}^p(X)$ are isomorphic even when V is not trivial. The point of this section is to prove this result, known as the *Thom isomorphism theorem*.

Definition 3.6.1 *Let V be a Hermitian vector bundle over a compact Hausdorff space X. We say that V is equipped with a fiberwise norm if:*

(i) *there exists a continuous function $\|\cdot\|$ from V to $[0, \infty)$ whose restriction to each fiber of V is a norm;*

(ii) *there exist constants c and C such that*

$$c \, \|v\|_{in} \leq \|v\| \leq C \, \|v\|_{in}$$

for all v in V.

Example 3.6.2 *If V is a Hermitian vector bundle over any compact Hausdorff space, then obviously the Hermitian metric defines a fiberwise norm on V.*

Example 3.6.3 *Let V and W be vector bundles over the same compact Hausdorff space, and suppose that V and W are each equipped with fiberwise norms $\|\cdot\|_V$ and $\|\cdot\|_W$ respectively. Then the function $\|\cdot\|$ on $V \oplus W$ defined by*

$$\|(v, w)\| = \sup\{\|v\|_V, \|w\|_W\}$$

is a fiberwise norm on $V \oplus W$; we can take $c = 1$ and $C = \sqrt{2}$.

There are many other examples, but we shall only need these two.

Definition 3.6.4 *Let V be a vector bundle over a compact Hausdorff space and suppose that V is equipped with a fiberwise norm $\|\cdot\|$. The*

ball bundle $B(V)$ *and the* sphere bundle $S(V)$ *of V with respect to* $\|\cdot\|$
are the sets

$$B(V) = \{v \in V : \|v\| \leq 1\}$$
$$S(V) = \{v \in V : \|v\| = 1\}.$$

Definition 3.6.5 *Let V be a vector bundle over a compact Hausdorff space X, let π be the projection map of V onto X, and suppose that V is equipped with a fiberwise norm. We construct a map*

$$K^0(X) \otimes K^0(B(V), S(V)) \longrightarrow K^0(B(V), S(V))$$

in the following way: define the compact pair morphism

$$p : (B(V), S(V)) \longrightarrow (X \times B(V), X \times S(V))$$

by the formula $p(v) = (\pi(v), v)$. Then compose the external multiplication map

$$K^0(X) \otimes K^0(B(V), S(V)) \longrightarrow K^0(X \times B(V), X \times S(V))$$

with the homomorphism

$$p^* : K^0(X \times B(V), X \times S(V)) \longrightarrow K^0(B(V), S(V)).$$

The function in Definition 3.6.5 makes $K^0(B(V), S(V))$ into a $K^0(X)$-module.

There is a related construction involving $K^{-1}(X)$ and $K^{-1}(V)$; first we need some preliminary results.

Lemma 3.6.6 *Let V be a vector bundle over a compact Hausdorff space and suppose that V is equipped with a fiberwise norm. Then $B(V)\backslash S(V)$ is homeomorphic to V.*

Proof The map $\phi : B(V)\backslash S(V) \longrightarrow V$ defined as $\phi(v) = (1 - \|v\|)^{-1}v$ is a homeomorphism. $\square$

Lemma 3.6.7 *Let V be a vector bundle over a compact Hausdorff space X and suppose that V is equipped with a fiberwise norm. Then there exists a natural isomorphism*

$$K^{-1}(B(V), S(V)) \cong K^0\big(B(V) \times S^1, (B(V) \times \{1\}) \cup (S(V) \times S^1)\big).$$

Proof We have homeomorphisms

$$(B(V) \times S^1) \backslash (B(V) \times \{1\}) \cup (S(V) \times S^1)) \cong$$
$$(B(V) \backslash S(V)) \times (S^1 \backslash \{1\}) \cong V \times \mathbb{R};$$

the second homeomorphism comes from Lemma 3.6.6. The result then follows from Theorem 2.2.14. □

Definition 3.6.8 *Let V be a vector bundle over a compact Hausdorff space X, let π be the projection map of V onto X, and suppose that V is equipped with a fiberwise norm. We construct a map*

$$K^{-1}(X) \otimes K^0(B(V), S(V)) \longrightarrow K^{-1}(B(V), S(V))$$

in the following way: Theorems 2.3.17 and 2.6.13 yield an isomorphism $K^{-1}(X) \cong K^0(X \times S^1, X \times \{1\})$. Apply external multiplication to produce a map

$$K^0(X \times S^1, X \times \{1\}) \otimes K^0(B(V), S(V)) \longrightarrow$$
$$K^0(X \times S^1 \times B(V), (X \times \{1\} \times B(V)) \cup (X \times S^1 \times S(V))).$$

Identify $X \times S^1 \times B(V)$ with $X \times B(V) \times S^1$ via the obvious homeomorphism, let p be as in Definition 3.6.5, and compose with

$$(p \times id)^* : K^0(X \times B(V) \times S^1, (X \times B(V) \times \{1\}) \cup (X \times S(V) \times S^1)) \longrightarrow$$
$$K^0(B(V) \times S^1, (B(V) \times \{1\} \cup (S(V) \times S^1))$$

to obtain a function

$$K^0(X \times S^1, X \times \{1\}) \otimes K^0(B(V), S(V)) \longrightarrow$$
$$K^0(B(V) \times S^1, (B(V) \times \{1\}) \cup (S(V) \times S^1)).$$

Finally, use Lemma 3.6.7 to make the identification

$$K^0(B(V) \times S^1, (B(V) \times \{1\}) \cup (S(V) \times S^1)) \cong K^{-1}(B(V), S(V)).$$

Suppose that $\pi : V \longrightarrow X$ is the projection map of a vector bundle. For each nonnegative integer k, we let $\wedge^k \pi$ denote the projection of the vector bundle $\bigwedge^k(V)$ to X, and we use Construction 1.5.12 to pull back $\bigwedge^k(V)$ to the vector bundle

$$\pi^*(\textstyle\bigwedge^k(V)) = \{(v, \omega) \in V \times \textstyle\bigwedge^k(V) : \pi(v) = (\wedge^k \pi)(\omega)\}$$

over V. This vector bundle is naturally isomorphic to $\bigwedge^k(\pi^* V)$, and for the sake of notational parsimony, we will use this latter notation for the bundle.

Lemma 3.6.9 *Let X be a compact Hausdorff space, suppose that V is a Hermitian vector bundle over X, and let k be a natural number. Define $\varepsilon : \bigwedge^k(\pi^*V) \longrightarrow \bigwedge^{k+1}(\pi^*V)$ by setting $\varepsilon(v, \omega) = (v, v \wedge \omega)$ for all v in V and ω in $\bigwedge^k(\pi^*V)$. For each element v of V that appears in a wedge product, let $\widehat{v}$ indicate omission of v. Then*

$$\varepsilon^*(v, v_1 \wedge \cdots \wedge v_k) = \left(v, \sum_{j=1}^{k+1} (-1)^{j+1} \langle v_j, v_1 \rangle v_1 \wedge \cdots \wedge \widehat{v}_j \wedge \cdots \wedge v_{k+1} \right)$$

for all simple wedges in $\bigwedge^k(V)_v$.

Proof For all simple wedges in $\bigwedge^k(V)_v$ and $\bigwedge^{k+1}(V)_v$, we have

$$\langle \varepsilon(v, v_1 \wedge \cdots \wedge v_k), v_1' \wedge \cdots \wedge v_{k+1}' \rangle = \langle v \wedge v_1 \wedge \cdots \wedge v_k, v_1' \wedge \cdots \wedge v_{k+1}' \rangle.$$

Let A denote the matrix whose $(1, j)$ entry is $\langle v, v_j' \rangle$ for $1 \leq j \leq k+1$ and whose (i, j) entry is $\{\langle v_{i-1}, v_j' \rangle\}$ when $1 < i \leq k+1$ and $1 \leq j \leq k+1$. Next, let $\mathsf{A}^{1,j}$ be the matrix obtained from A by removing the first row and the jth column. Then Definition 1.1.1, Proposition 3.5.7, and properties of the determinant give us

$$\langle \varepsilon(v, v_1 \wedge \cdots \wedge v_{k+1}), v_1' \wedge \cdots \wedge v_{k+1}' \rangle = \det \mathsf{A}$$

$$= \sum_{j=1}^{k+1} (-1)^{j+1} \langle v, v_j' \rangle \det(\mathsf{A}^{1,m})$$

$$= \sum_{j=1}^{k+1} (-1)^{j+1} \langle v, v_j' \rangle \left\langle v \wedge v_1 \wedge \cdots \wedge v_{k+1}, v_1' \wedge \cdots \wedge \widehat{v'}_j \wedge \cdots \wedge v_{k+1}' \right\rangle$$

$$= \left\langle v_1 \wedge \cdots \wedge v_{k+1}, \sum_{j=1}^{k+1} (-1)^{j+1} \langle v_j', v \rangle v_1' \wedge \cdots \wedge \widehat{v'}_j \wedge \cdots \wedge v_{k+1}' \right\rangle,$$

from which the desired formula follows. $\square$

Proposition 3.6.10 *Let V and W be vector bundles over a compact Hausdorff space X. There exists a natural bundle isomorphism*

$$\bigwedge^n \left(\pi_{V \oplus W}^*(V \oplus W) \right) \cong \bigoplus_{k+l=n} \bigwedge^k(\pi_V^* V) \boxtimes \bigwedge^l(\pi_W^* W)$$

for every nonnegative integer n. Via this isomorphism, the map

$$\varepsilon : \bigwedge^n(\pi_{V \oplus W}^*(V \oplus W)) \longrightarrow \bigwedge^{n+1}(\pi_{V \oplus W}^*(V \oplus W))$$

takes the form

$$\varepsilon : \bigwedge^k(\pi_V^* V) \boxtimes \bigwedge^l(\pi_W^* W) \longrightarrow$$
$$\left(\bigwedge^{k+1}(\pi_V^* V) \boxtimes \bigwedge^l(\pi_W^* W)\right) \oplus \left(\bigwedge^k(\pi_V^* V) \boxtimes \bigwedge^{l+1}(\pi_W^* W)\right)$$

for all k and l summing to n, and

$$\varepsilon_{V \oplus W} = \varepsilon_V \boxtimes 1 + (-1)^k \boxtimes \varepsilon_W.$$

In addition, suppose that V and W are Hermitian bundles and endow $V \oplus W$ with the product Hermitian metric; i.e, $\langle (v_1, w_1), (v_2, w_2) \rangle = \langle v_1, v_2 \rangle + \langle w_1, w_2 \rangle$. Then

$$\varepsilon_{V \oplus W}^* = \varepsilon_V^* \boxtimes 1 + (-1)^k \boxtimes \varepsilon_W^*.$$

Proof By applying Proposition 3.5.6, we obtain a bijection

$$\tau : \bigoplus_{k+l=n} \bigwedge^k(\pi_V^* V) \boxtimes \bigwedge^l(\pi_W^* W) \longrightarrow \bigwedge^n \left(\pi_{V \oplus W}^*(V \oplus W)\right)$$

that maps fibers to fibers. Local triviality of vector bundles and the formula for τ in Proposition 3.5.6 imply that τ is continuous, and so τ is a bundle isomorphism by Lemma 1.7.8.

To verify the alleged formula for ε, we only need to check it on simple wedges. Take $(v, v_1 \wedge \cdots \wedge v_k)$ and $(w, w_1 \wedge \cdots \wedge w_l)$ in $\bigwedge^k(\pi_V^* V)$ and $\bigwedge^l(\pi_W^* W)$ respectively. Then

$$\varepsilon_{V \oplus W} \tau \big((v, v_1 \wedge \cdots \wedge v_k) \otimes (w, w_1 \wedge \cdots \wedge w_l)\big)$$
$$= \varepsilon_{V \oplus W} \big((v, w), (v_1, 0) \wedge \cdots \wedge (v_k, 0) \wedge (0, w_1) \wedge \cdots \wedge (0, w_l)\big)$$
$$= \big((v, w), (v, w) \wedge (v_1, 0) \wedge \cdots \wedge (v_k, 0) \wedge (0, w_1) \wedge \cdots \wedge (0, w_l)\big)$$
$$= \big((v, w), ((v, 0) + (0, w)) \wedge (v_1, 0) \wedge \cdots \wedge (v_k, 0) \wedge (0, w_1) \wedge \cdots \wedge (0, w_l)\big)$$
$$= \big((v, w), (v, 0) \wedge (v_1, 0) \wedge \cdots \wedge (v_k, 0) \wedge (0, w_1) \wedge \cdots \wedge (0, w_l)\big)$$
$$+ (-1)^k \big((v, w), (v_1, 0) \wedge \cdots \wedge (v_k, 0) \wedge (0, w) \wedge (0, w_1) \wedge \cdots \wedge (0, w_l)\big),$$

whence $\varepsilon_{V \oplus W} = \varepsilon_V \boxtimes 1 + (-1)^k \boxtimes \varepsilon_W$. An easy computation shows that the desired formula for $\varepsilon_{V \oplus W}^*$ holds as well. $\qquad\square$

Proposition 3.6.11 *Let V be a Hermitian vector bundle over a compact Hausdorff space X and equip V with any fiberwise norm. Then*

$$\left(\bigwedge^{even}(\pi^* V), \bigwedge^{odd}(\pi^* V), \varepsilon + \varepsilon^*\right)$$

determines an element $\mathbf{U}_V$ of $K^0(B(V), S(V))$.

Proof The vector bundles above are restricted to $B(V)$, but we omit that dependence in order to simplify the notation. Local triviality and our formulas for ε and ε^* imply that both maps are continuous. To show that $\varepsilon + \varepsilon^*$ is a bundle isomorphism, it suffices to show that its square is a bundle isomorphism. Proposition 3.5.4 and the definition of ε immediately give us $\varepsilon^2 = 0$, and thus $(\varepsilon^*)^2 = 0$ by Proposition 1.1.9. Therefore $(\varepsilon + \varepsilon^*)^2 = \varepsilon\varepsilon^* + \varepsilon^*\varepsilon$. A straightforward computation then yields

$$(\varepsilon^*\varepsilon + \varepsilon\varepsilon^*)(v, v_1 \wedge v_2 \wedge \cdots \wedge v_{k+1}) = (v, \langle v_1, v_1 \rangle v_2 \wedge v_3 \wedge \cdots \wedge v_{k+1}).$$

The vector v is not zero and hence $(\varepsilon + \varepsilon^*)^2$ is the identity map times a strictly positive function on $S(V)$, implying the conclusion of the proposition. □

Definition 3.6.12 *The element* $\mathbf{U}_V$ *constructed in Proposition 3.6.11 is called the* Thom class *of V.*

Lemma 3.6.6 and the following result justify calling $\mathbf{U}_V$ the Thom class of V and not the Thom class of $(B(V), S(V))$.

Proposition 3.6.13 *Let V be a vector bundle over a compact Hausdorff space X. For $k = 0, 1$, let V_k denote V equipped with a Hermitian structure $\langle \cdot, \cdot \rangle_k$ and a fiberwise norm $\|\cdot\|_k$ Then there exists an morphism ι from $(B(V_0), S(V_0))$ to $(B(V_1), S(V_1))$ with the properties that*

$$\iota^* : \mathrm{K}^0(B(V_1), S(V_1)) \longrightarrow \mathrm{K}^0(B(V_0), S(V_0))$$

is an isomorphism and that $\iota^*(\mathbf{U}_{V_1}) = \mathbf{U}_{V_0}$ in $\mathrm{K}^0(B(V_0), S(V_0))$.

Proof We first consider the case where the Hermitian structure changes but the fiberwise norm does not, and second the case where the fiberwise norm changes but the Hermitian structure remains fixed; the general result follows easily from these two special cases.

In the first case, we have $B(V_0) = B(V_1)$ and $S(V_0) = S(V_1)$, so we can take ι to be the identity map. For each $0 \leq t \leq 1$, define a Hermitian metric on V by the formula

$$\langle v, v' \rangle_t = (1 - t) \langle v, v' \rangle_0 + t \langle v, v' \rangle_1.$$

The map ε in the definition of the Thom class does not depend on the Hermitian structure, while ε^* varies continuously in t. Thus $\mathbf{U}_{V_0} = \mathbf{U}_{V_1}$ by Proposition 3.4.11.

In the second case, $\varepsilon + \varepsilon^*$ does not change, but the ball and sphere bundles of V do. Define $\iota : (B(V_0), S(V_0)) \longrightarrow (B(V_1), S(V_1))$ by the formula

$$\iota(v) = \begin{cases} \frac{\|v\|_0}{\|v\|_1} v & v \neq 0 \\ 0 & v = 0. \end{cases}$$

The definition of fiberwise norm implies that ι is continuous. Moreover, if we define ι' by reversing the roles of $(B(V_0), S(V_0))$ and $(B(V_1), S(V_1))$ in the definition of ι, then the compositions $\iota'\iota$ and $\iota\iota'$ are the identity maps on $B(V_0)$ and $B(V_1)$ respectively, whence ι^* is an isomorphism. Finally, an easy check shows that $\iota^*(\mathbf{U}_{V_1}) = \mathbf{U}_{V_0}$. □

Proposition 3.6.14 *Let V and W be Hermitian vector bundles over a compact Hausdorff space X, each endowed with the fiberwise norm coming from the Hermitian structure. Endow $V \oplus W$ with the fiberwise norm*

$$\|(v, w)\| = \sup\{\|v\|_{in}, \|w\|_{in}\}.$$

Then

$$B(V \oplus W) = B(V) \times B(W)$$
$$S(V \oplus W) = \big(B(V) \times S(W)\big) \cup \big(S(V) \times B(W)\big)$$

and $\mathbf{U}_{V \oplus W} = \mathbf{U}_V \boxtimes \mathbf{U}_W$.

Proof The facts about $B(V \oplus W)$ and $S(V \oplus W)$ follow directly from the definitions of the fiberwise norms; Theorem 3.4.17 and Proposition 3.6.10 imply the last equality. □

In the lemma and theorem that follow, we denote the operations defined in Definitions 3.6.5 and 3.6.8 by a dot.

Lemma 3.6.15 *For every compact Hausdorff space X and every natural number n, the map*

$$(\ )\cdot \mathbf{U}_{\Theta^n(X)} : \mathrm{K}^\mathrm{P}(X) \longrightarrow \mathrm{K}^\mathrm{P}\big(B(\Theta^n(X)), S(\Theta^n(X))\big)$$

is an isomorphism for $p = 0, -1$.

Proof We proceed by induction on n; to simplify notation we shorten $\Theta^n(X)$ to just Θ^n. Proposition 3.6.13 implies that our choice of fiberwise norm on each Θ^n is not important; for $n = 1$, we choose the fiberwise norm that comes from the standard Hermitian metric on Θ^1, and thus $B(\Theta^1) = B^2$ and $S(\Theta^1) = S^1$. For $n > 1$, we choose the product fiberwise norm.

To minimize confusion, we denote Θ^1 by V and let π be the projection map to X. Because π^*V is a line bundle, we have

$$\bigwedge^{even}(\pi^*V) = \bigwedge^0(\pi^*V) = \Theta^1(B(V))$$

and

$$\bigwedge^{odd}(\pi^*V) = \bigwedge^1(\pi^*V) = \pi^*V.$$

In this case $\varepsilon(v, z) = vz$ for every v in $B(V)$ and z in $\mathbb{C}$, while ε^* is identically 0. Thus

$$[W] \cdot \mathbf{U}_V = [p^*(W \boxtimes \Theta^1(B(V))), p^*(W \boxtimes \pi^*V), p^*(I \boxtimes \varepsilon)]$$

for every bundle W over X. Because V is a trivial line bundle, the space $B(V))$ is homeomorphic to $X \times B^2$, and

$$p^*(W \boxtimes \pi^*V)_{(x,b)} = \left\{ \left(x, b, \sum_i w_i \otimes (x, b, v_i) \right) : w_i \in W, v_i \in V \right\}$$

for each point (x, b) in $X \times B^2$. We also have

$$(W \otimes \pi^*V)_{(x,b)} = \left\{ \sum_i w_i \otimes (x, b, v_i) : w_i \in W, v_i \in V \right\},$$

and hence

$$p^*(W \boxtimes \pi^*V) \cong W \boxtimes \pi^*V.$$

A similar argument yields

$$p^*(W \boxtimes \Theta^1(B(V))) \cong W \boxtimes \Theta^1(B(V)).$$

Therefore, using the notation of Proposition 3.3.9, we obtain

$$\begin{aligned}
[W] \cdot \mathbf{U}_V &= [p^*(W \boxtimes \Theta^1(B(V))), p^*(W \boxtimes \pi^*V), p^*(I \boxtimes \varepsilon)] \\
&= [W \boxtimes \Theta^1(B(V)), W \boxtimes \pi^*V, \mathrm{id} \boxtimes \varepsilon] \\
&= [W] \boxtimes \tilde{\mathbf{b}}.
\end{aligned}$$

Proposition 3.3.9 then implies that the map defined as

$$\mathbf{x} \mapsto \mathbf{x} \cdot \mathbf{U}_{\Theta^1(X)}$$

for all $\mathbf{x}$ in $\mathrm{K}^0(X)$ determines an isomorphism

$$\mathrm{K}^0(X) \cong \mathrm{K}^0\big(B(\Theta^1), S(\Theta^1)\big).$$

Similarly, take $[W_0, W_1, \sigma]$ in $\mathrm{K}^0(X \times S^1, X \times \{1\})$, let $\tilde{\sigma}$ be an extension of σ to a bundle homomorphism from W_0 to W_1, and define a matrix

$$\gamma = \begin{pmatrix} \tilde{\sigma} \boxtimes I & 0 \\ I \boxtimes \varepsilon & \tilde{\sigma}^* \boxtimes I \end{pmatrix}.$$

Then Theorem 3.4.17 yields

$$[W_0, W_1, \sigma] \cdot \mathbf{U}_V$$
$$= \big((W_0 \boxtimes \Theta^1(B(V)) \oplus (W_1 \boxtimes \pi^*V), (W_1 \boxtimes \Theta^1(B(V)) \oplus (W_0 \boxtimes \pi^*V), \gamma\big)$$
$$= [W_0, W_1, \sigma] \boxtimes \tilde{\mathbf{b}},$$

and therefore we have an isomorphism $\mathrm{K}^{-1}(X) \cong \mathrm{K}^{-1}\big(B(\Theta^1), S(\Theta^1)\big)$ as well.

Now suppose the lemma is true for Θ^n. Proposition 3.6.14 implies that the diagram

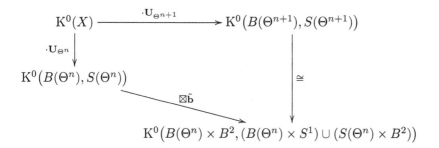

is commutative, and thus $\mathrm{K}^0(X) \cong \mathrm{K}^0\big(B(\Theta^{n+1}), S(\Theta^{n+1})\big)$.

Finally, define groups

$$\mathcal{G}_1 = \mathrm{K}^0\big(B(\Theta^n) \times S^1 \times B^2,$$
$$(B(\Theta^n) \times S^1 \times S^1) \cup (B(\Theta^n) \times \{1\} \times B^2) \cup (S(\Theta^n) \times S^1 \times B^2)\big)$$
$$\mathcal{G}_2 = \mathrm{K}^0\big(B(\Theta^n) \times B^2 \times S^1,$$
$$(B(\Theta^n) \times S^1 \times S^1) \cup (B(\Theta^n) \times B^2 \times B^2) \cup (S(\Theta^n) \times B^2 \times S^1)\big).$$

Let $\gamma : B(\Theta^n) \times B^2 \times S^1 \longrightarrow B(\Theta^n) \times S^1 \times B^2$ be the map that swaps the second and third factors. Then $\gamma^* : \mathcal{G}_1 \longrightarrow \mathcal{G}_2$ is an isomorphism,

and the diagram

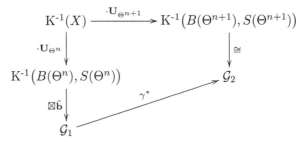

commutes by Definition 3.6.8 and Proposition 3.6.14. Therefore the Thom class of Θ^{n+1} determines a group isomorphism from $K^{-1}(X)$ to $K^{-1}\big(B(\Theta^{n+1}), S(\Theta^{n+1})\big)$.   □

Theorem 3.6.16 (Thom isomorphism theorem) *If V is a vector bundle over a compact Hausdorff space X, then the Thom class $\mathbf{U}_V$ of V defines group isomorphisms $K^p(X) \cong K^p(B(V), S(V))$ for $p = 0, -1$.*

Proof Suppose we have a finite collection $\mathcal{A}$ of closed subsets of X whose union is X and such that V is trivial when restricted to each set in $\mathcal{A}$. We prove the theorem by induction on the number of sets in $\mathcal{A}$. If $\mathcal{A}$ consists of only one set, i.e., X, then the desired result follows immediately from Lemma 3.6.15. Now suppose the theorem is true when $\mathcal{A}$ has n sets, and consider the commutative diagram

$$
\begin{array}{ccc}
K^p(A_1) \oplus K^p(A) & \xrightarrow{\;\cdot(\mathbf{U}_{V|A_1} \oplus \mathbf{U}_{V|A})\;} & K^p(B(V), S(V)) \\
\big\downarrow & & \big\downarrow \\
K^p(A_1 \cap A) & \xrightarrow{\;\cdot\mathbf{U}_{V|A_1 \cap A}\;} & K^p(B(V|A_1 \cap A), S(V|A_1 \cap A)) \\
\big\downarrow & & \big\downarrow \\
K^p(X) & \xrightarrow{\;\cdot\mathbf{U}_V\;} & K^p(B(V), S(V)) \\
\big\downarrow & & \big\downarrow \\
K^{-(p+1)}(A_1) \oplus K^{-(p+1)}(A) & \xrightarrow{\;\cdot(\mathbf{U}_{V|A_1} \oplus \mathbf{U}_{V|A})\;} & K^{-(p+1)}(B(V), S(V)) \\
\big\downarrow & & \big\downarrow \\
K^{-(p+1)}(A_1 \cap A) & \xrightarrow{\;\cdot\mathbf{U}_{V|A_1 \cap A}\;} & K^{-(p+1)}(B(V|A_1 \cap A), S(V|A_1 \cap A)),
\end{array}
$$

where $\mathcal{A} = \{A_1, A_2, \ldots, A_n, A_{n+1}\}$ and $A = A_2 \cup A_3 \cup \cdots \cup A_n \cup A_{n+1}$. The first column is exact by Theorem 3.1.2. Moreover, Lemma 3.6.6 provides a homeomorphism $B(V|\widetilde{A}) \backslash S(V|\widetilde{A}) \cong V|\widetilde{A}$ for every closed subset $\widetilde{A}$ of X, so Theorem 3.1.2 also tells us that the second column is exact. Each of the horizontal maps except the middle one is an isomorphism, either by the inductive hypothesis or Lemma 3.6.15, and the naturality of all the relevant maps implies that the diagram commutes. The theorem then is a consequence of Proposition 1.8.10. $\qquad\square$

3.7 The splitting principle

The *splitting principle* states, roughly speaking, that we can assume in many applications of K-theory that every vector bundle is an internal Whitney sum of line bundles. In proving the splitting principle, we will obtain along the way some computations of the K^0 groups of the complex projective spaces $\mathbb{CP}^n$ for all natural numbers n.

In Example 1.5.10 we defined the tautological line bundle over complex projective space $\mathbb{CP}^n$. We now generalize that construction.

Definition 3.7.1 *Let V be a vector bundle over a compact Hausdorff space X. For each point x in X, identify x with the zero element in V_x; this defines an imbedding of X into V. Identify v with λv for every complex number λ. The resulting quotient of $V \backslash X$ is denoted $\mathbb{P}(V)$ and is called the* projective bundle *associated to V.*

For each element ℓ of $\mathbb{P}(V)$, there is some x in X such that ℓ can be viewed a line in V_x that passes through the origin; i.e., a one-dimensional vector subspace of V_x.

Note that $\mathbb{P}(V)$ is *not* a vector bundle.

Definition 3.7.2 *Let V be a vector bundle over a compact Hausdorff space X and let $p : \mathbb{P}(V) \longrightarrow X$ be the continuous function determined by the projection from V to X. The* dual Hopf bundle *associated to V is the subbundle*

$$H_V^* = \{(\ell, v) \in \mathbb{P}(V) \oplus V : \ell \text{ contains } v\}$$

of the vector bundle $p^*(V)$ over $\mathbb{P}(V)$.

Definition 3.7.3 *Let V be a vector bundle over a compact Hausdorff space X. The* Hopf bundle H_V *associated to V is the vector bundle $(H_V^*)^*$ over $\mathbb{P}(V)$.*

When the vector bundle V is clear from the context, we will write H_V as just H.

Proposition 3.7.4 *Let n be a natural number, suppose that V is a rank n vector bundle over a compact Hausdorff space X, and let H denote the Hopf bundle associated to V. Then in $K^0(\mathbb{P}(V))$ we have the equation*

$$\sum_{k=0}^{n}(-1)^k\left[\textstyle\bigwedge^k(p^*V)\right]\otimes[H]^k = 0.$$

Proof Let $p : \mathbb{P}(V) \longrightarrow X$ be as in Definition 3.7.2 and use Proposition 1.7.15 to construct a vector bundle W such that $W \oplus H^*$ is (isomorphic to) p^*V. From Proposition 3.5.11 and the fact that H^* is a line bundle we obtain

$$\textstyle\bigwedge_t[(p^*V)] = \bigwedge_t([W]) \otimes \bigwedge_t([H^*]) = \bigwedge_t([W]) \otimes \left([\Theta^1(X)] + [H^*]t\right).$$

If we evaluate this last factor at $t = -[H]$, Proposition 3.3.5 gives us

$$[\Theta^1(X)] - \left([H] \otimes [H^*]\right) = [\Theta^1(X)] - [\Theta^1(X)] = 0,$$

and therefore

$$0 = \textstyle\bigwedge_{-[H]}[(p^*V)] = \sum_{k=0}^{n}(-1)^k\left[\bigwedge^k(p^*V)\right]\otimes[H]^k.$$

$\square$

Lemma 3.7.5 *Let X and Y be compact Hausdorff spaces and suppose that $\phi : Y \longrightarrow X$ is continuous. Choose elements $\mathbf{y}_1, \mathbf{y}_2, \ldots, \mathbf{y}_n$ of $K^0(Y)$, let $\mathcal{G}$ be the free abelian group generated by these elements, and let $\phi : \mathcal{G} \longrightarrow K^0(Y)$ be the quotient homomorphism. Suppose that each point x in X has a neighborhood U with the feature that for each closed subspace A' of $\overline{U}$, with $j : \phi^{-1}(A') \longrightarrow Y$ denoting inclusion, the homomorphism $\Phi' : \mathcal{G} \otimes K^p(A') \longrightarrow K^p(\phi^{-1}(A'))$ defined by the formula*

$$\Phi'\left(\sum_i g_i \otimes \mathbf{a}_i\right) = \sum_i j^*\rho(g_1) \otimes \left(\phi|\phi^{-1}(A')\right)^*(\mathbf{a}_i)$$

is an isomorphism for $p = 0, -1$. Then for every closed subspace A of X, the map Φ' determines an isomorphism Φ from $\mathcal{G} \otimes K^p(X, A)$ to $K^p(Y, \phi^{-1}(A))$. In particular, there exists an isomorphism

$$\Phi : \mathcal{G} \otimes K^p(X) \longrightarrow K^p(Y).$$

Proof For every compact pair (X, A), we have a diagram

$$
\begin{array}{ccccccc}
0 & \longrightarrow & \mathcal{G} \otimes \mathrm{K}^\mathrm{p}(X, A) & \longrightarrow & \mathcal{G} \otimes \mathrm{K}^\mathrm{p}(\mathcal{D}(X, A)) & \longrightarrow & \mathcal{G} \otimes \mathrm{K}^\mathrm{p}(X) \\
& & \downarrow & & {\scriptstyle \Phi'}\downarrow & & {\scriptstyle \Phi'}\downarrow \\
0 & \longrightarrow & \mathrm{K}^\mathrm{p}(Y, \phi^{-1}(A)) & \longrightarrow & \mathrm{K}^\mathrm{p}(\mathcal{D}(Y, \phi^{-1}(A))) & \longrightarrow & \mathrm{K}^\mathrm{p}(Y)
\end{array}
$$

for $p = 0, -1$. Theorem 2.2.10, Theorem 2.3.15, Proposition 3.2.7, and Definition 2.2.2 imply that the rows are exact, and then Proposition 1.8.9 gives us Φ as the leftmost map in the diagram.

Let x_1 and x_2 be distinct points in X and choose neighborhoods U_1 and U_2 of x_1 and x_2 respectively that satisfy the hypotheses of the lemma. Let $U = U_1 \cup U_2$, suppose that A is a closed subspace of $\overline{U}$, and set

$$
\begin{aligned}
B &= \overline{U}_1 \cup A \\
A_1 &= A \cap \overline{U}_1 \\
A_2 &= B \cap \overline{U}_2.
\end{aligned}
$$

From Theorem 2.7.17 and the fact that $\overline{U} \backslash B = \big((\overline{U} \backslash A) \backslash (B \backslash A)\big)$ we obtain the six-term exact sequence

$$
\begin{array}{ccccc}
\mathrm{K}^0(\overline{U} \backslash B) & \longrightarrow & \mathrm{K}^0(\overline{U} \backslash A) & \longrightarrow & \mathrm{K}^0(B \backslash A) \\
\uparrow & & & & \downarrow \\
\mathrm{K}^{-1}(\overline{U} \backslash B) & \longleftarrow & \mathrm{K}^{-1}(\overline{U} \backslash A) & \longleftarrow & \mathrm{K}^{-1}(\overline{U} \backslash B).
\end{array}
$$

But $\overline{U} \backslash B = \overline{U}_2 \backslash A_2$ and $B \backslash A = \overline{U}_1 \backslash A_1$, so Proposition 2.6.8 implies that the sequence

$$
\begin{array}{ccccc}
\mathrm{K}^0(\overline{U}_2, A_2) & \longrightarrow & \mathrm{K}^0(\overline{U}, A) & \longrightarrow & \mathrm{K}^0(\overline{U}_1, A_1) \\
\uparrow & & & & \downarrow \\
\mathrm{K}^{-1}(\overline{U}_1, A_1) & \longleftarrow & \mathrm{K}^{-1}(\overline{U}, A) & \longleftarrow & \mathrm{K}^{-1}(\overline{U}_2, A_2)
\end{array}
$$

is exact. We get a second exact sequence by taking by taking the inverse image under ϕ of each of the spaces that appear in the above exact sequence. Proposition 3.2.7 then implies that for $p = 0, -1$, we have a

diagram

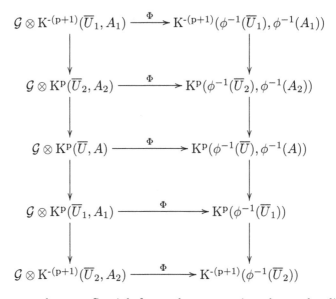

with exact columns. Straightforward computation shows the diagram commutes, and thus Proposition 1.8.10 implies that

$$\Phi : \mathcal{G} \otimes \mathrm{K}^{\mathrm{p}}(\overline{U}, A) \longrightarrow \mathrm{K}^{\mathrm{p}}(\phi^{-1}(\overline{U}), \phi^{-1}(A))$$

is an isomorphism for $p = 0, -1$. The general result follows by mathematical induction on the number of sets in a finite open cover of X. $\qquad\square$

Proposition 3.7.6 *Let V be a line bundle over a compact Hausdorff space X and let H be the Hopf bundle associated to $V \oplus \Theta^1(X)$. Then $\mathrm{K}^0(\mathbb{P}(V \oplus \Theta^1(X)))$ is a free $\mathrm{K}^0(X)$-module on the generators $[H]$ and $[\Theta^1(\mathbb{P}(V \oplus \Theta^1(X)))]$. Furthermore, we have the equation*

$$\big([H] \cdot [V] - [\Theta^1(\mathbb{P}(V \oplus \Theta^1(X)))]\big) \otimes \big([H] - [\Theta^1(\mathbb{P}(V \oplus \Theta^1(X)))]\big) = 0.$$

Proof To simplify notation in this proof, we write $Y = \mathbb{P}(V \oplus \Theta^1(X))$.

Let $p : Y \longrightarrow X$ be the projection map. We make $\mathrm{K}^0(Y)$ into a $\mathrm{K}^0(X)$-module by applying p^* to elements of $\mathrm{K}^0(X)$ and then performing internal multiplication with elements of $\mathrm{K}^0(Y)$.

Let U be an open subspace of X such that $V|\overline{U}$ is trivial. Then

$$\mathbb{P}(V|\overline{U} \oplus \Theta^1(\overline{U})) \cong \mathbb{P}(\Theta^1(\overline{U}) \oplus \Theta^1(\overline{U})) \cong \overline{U} \times \mathbb{CP}^1,$$

and Proposition 3.3.15 gives us

$$K^0(\overline{U} \times \mathbb{CP}^1) \cong \left(K^0(\overline{U}) \otimes K^0(\mathbb{CP}^1)\right) \oplus \left(K^{-1}(\overline{U}) \otimes K^{-1}(\mathbb{CP}^1)\right)$$
$$= K^0(\overline{U}) \otimes K^0(\mathbb{CP}^1).$$

Example 2.8.2 implies that $K^0(\mathbb{CP}^1)$ is generated by $[\Theta^1(\mathbb{CP}^1)]$ and the dual Hopf bundle over $\mathbb{CP}^1$, and because tensor product with $[H]$ is a bijection, we see that $K^0(\mathbb{CP}^1)$ is alternately generated by $[\Theta^1(\mathbb{CP}^1)$ and $[H]$. Therefore Proposition 3.3.15 implies that $K^0(\overline{U} \times \mathbb{CP}^1)$ is generated by the trivial bundle and the Hopf bundle over $\overline{U} \times \mathbb{CP}^1$.

Now consider the trivial bundle $\Theta^1(Y)$. Because H restricts to the Hopf bundle over $\mathbb{P}(\overline{U} \oplus \Theta^1(\overline{U}))$ for every open subspace U of X, the hypothesis of Lemma 3.7.5 are satisfied. Thus if $\mathcal{G}$ is the free abelian group generated by $[\Theta^1(Y)]$ and $[H]$, we have an isomorphism

$$\mathcal{G} \otimes K^0(X) \cong K^0(Y).$$

In other words, every element of $K^0(Y)$ can be uniquely written as a linear combination of $[\Theta^1(Y)]$ and $[H]$ with coefficients from $K^0(X)$. This proves the first statement of the proposition.

To establish the second statement, note that because H and V are line bundles, Proposition 3.5.11 implies that

$$\textstyle\bigwedge^2 \left(p^*(V \oplus \Theta^1(X))\right) \cong \bigwedge^1(p^*V) \otimes \bigwedge^1(\Theta^1(Y)) \cong p^*V \otimes \Theta^1(Y) \cong p^*V.$$

Thus Proposition 3.7.4 yields

$$\begin{aligned}
0 &= [\Theta^1(Y)] - [p^*(V \oplus \Theta^1(X))] \otimes [H] + [p^*V] \otimes [H]^2 \\
&= [\Theta^1(Y)] - \left(([p^*V] + [\Theta^1(Y)]) \otimes [H] + [p^*V] \otimes [H]^2\right) \\
&= \left([H] \otimes [p^*V] - [\Theta^1(Y)]\right) \otimes \left([H] - [\Theta^1(Y)]\right) \\
&= \left([H] \cdot [V] - [\Theta^1(Y)]\right) \otimes \left([H] - [\Theta^1(Y)]\right).
\end{aligned}$$

$\square$

Proposition 3.7.7 *Let n be a natural number and let H be the Hopf bundle over $\mathbb{P}(\Theta^{n+1}(pt)) \cong \mathbb{CP}^n$. Then $K^0(\mathbb{CP}^n)$ is a free abelian group on generators $[\Theta^1(\mathbb{CP}^n)], [H], [H]^2, \ldots, [H]^n$. Furthermore, we have the equation*

$$\left([H] - [\Theta^1(\mathbb{CP}^n)]\right)^{n+1} = 0.$$

Proof Apply Proposition 3.7.4 to the vector bundle $\Theta^{n+1}(pt)$ and use

Proposition 3.5.5 to obtain the string of equalities

$$0 = \sum_{k=1}^{n+1}(-1)^k[\textstyle\bigwedge^k(p^*\Theta^{n+1}(pt))] \otimes [H]^k$$

$$= \sum_{k=1}^{n+1}(-1)^k[\textstyle\bigwedge^k(\Theta^{n+1}(\mathbb{CP}^n))] \otimes [H]^k$$

$$= \sum_{k=1}^{n+1}(-1)^k\binom{n+1}{k}[\Theta^1(\mathbb{CP}^n)] \otimes [H]^k$$

$$= \sum_{k=1}^{n+1}(-1)^k\binom{n+1}{k}[\Theta^1(\mathbb{CP}^n)]^{n+1-k} \otimes [H]^k$$

$$= ([\Theta^1(\mathbb{CP}^n)] - [H])^{n+1}$$

$$= ([H] - [\Theta^1(\mathbb{CP}^n)])^{n+1}.$$

To establish the first part of the theorem, we proceed by induction. For $n = 1$, the result is a consequence of Proposition 3.7.6. Now suppose the result is true for an arbitrary n. To avoid confusion, we will denote the Hopf bundle of $\mathbb{CP}^n$ by H_n.

Consider the vector bundle $V = H_n^* \oplus \Theta^1(\mathbb{CP}^n)$ over $\mathbb{CP}^n$. Let p be the projection map from $\mathbb{P}(V)$ to $\mathbb{CP}^n$, and define $s : \mathbb{CP}^n \longrightarrow \mathbb{P}(V)$ by sending each x in $\mathbb{CP}^n$ to the line through the point $(0, x, 1)$. Each element of H_n^* is a line in $\mathbb{C}^{n+1}$, and so we can define a continuous function $\tilde{q} : \mathbb{P}(V) \longrightarrow \mathbb{CP}^{n+1}$ by sending the line through a point (h, x, z) to the line through (h, z). The map $\tilde{q}$ determines a homeomorphism

$$q : (\mathbb{P}(V) \backslash s(\mathbb{CP}^n))^+ \longrightarrow \mathbb{CP}^{n+1}.$$

Furthermore, if $\widehat{H}$ is the Hopf bundle over $\mathbb{P}(V)$, then $q^*H_{n+1} \cong \widehat{H}$ and $s^*\widehat{H} \cong \Theta^1(\mathbb{CP}^1)$.

Next, we have a split exact sequence

$$0 \longrightarrow \mathrm{K}^0\big(\mathbb{P}(V), s(\mathbb{CP}^n)\big) \longrightarrow \mathrm{K}^0(\mathbb{P}(V)) \xrightarrow{\quad\longleftarrow\quad} \mathrm{K}^0(s(\mathbb{CP}^n)) \longrightarrow 0;$$

the splitting map takes the line through a point (h, x, z) to the line through the point $(0, x, z)$. Theorem 2.2.10 gives us a group isomorphism

$$\mathrm{K}^0(\mathbb{P}(V)) \cong \mathrm{K}^0\big(\mathbb{P}, s(\mathbb{CP}^n)\big) \oplus \mathrm{K}^0(\mathbb{CP}^n).$$

By Proposition 3.7.6, the group $\mathrm{K}^0(\mathbb{P}(V))$ is a free $\mathrm{K}^0(\mathbb{CP}^n)$-module on the generators $[\Theta^1(\mathbb{P}(V))]$ and $[\widehat{H}]$. Combining the exact sequence above with our inductive hypothesis, we see that every element of the group $K^0\big(\mathbb{P}(V), s(\mathbb{CP}^n)\big)$ can be uniquely written in the form

$$\left(\sum_{i=0}^{n} k_i [H_n]^i\right) \cdot \big([\widehat{H}] - [\Theta^1(\mathbb{P}(V))]\big)$$

for some integers $k_0, k_1, \ldots, k_n$. Next, apply Proposition 3.7.6 to the line bundle H_n^* to obtain the equation

$$\big([\widehat{H}] \cdot [H_n^*] - [\Theta^1(\mathbb{P}(V))]\big) \otimes \big([\widehat{H}] - [\Theta^1(\mathbb{P}(V))]\big) = 0,$$

which from the definition of the $K^0(\mathbb{CP}^n)$-module becomes

$$\big([\widehat{H}] \otimes [p^* H_n^*] - [\Theta^1(\mathbb{P}(V))]\big) \otimes \big([\widehat{H}] - [\Theta^1(\mathbb{P}(V))]\big) = 0.$$

Tensor both sides of the equation by $[H_n] \otimes [\Theta^1(\mathbb{P}(V))]$ to obtain

$$\big([\widehat{H}] - [p^* H_n]\big) \otimes \big([\widehat{H}] - [\Theta^1(\mathbb{P}(V))]\big) = 0,$$

which can be rewritten in the form

$$[\widehat{H}] \otimes \big([\widehat{H}] - [\Theta^1(\mathbb{P}(V))]\big) = [H_n] \cdot \big([\widehat{H}] - [\Theta^1(\mathbb{P}(V))]\big).$$

Thus the set

$$\big\{[\widehat{H}]^i \cdot \big([\widehat{H}] - [\Theta^1(\mathbb{P}(V))]\big) : 0 \le i \le n\big\}$$

is a basis for $K^0\big(\mathbb{P}(V), s(\mathbb{CP}^n)\big)$, and via Theorem 2.2.14 and the bundle isomorphism $q^* H_{n+1} \cong \widehat{H}$, we see that

$$\big\{[H_{n+1}]^i \cdot \big([H_{n+1}] - [\Theta^1(\mathbb{CP}^{n+1})]\big) : 0 \le i \le n\big\}$$

is a basis for the group $\widetilde{K}^0(\mathbb{CP}^{n+1})$. Finally, we combine this fact with Corollary 2.5.4 to conclude that $K^0(\mathbb{CP}^{n+1})$ is a free abelian group on generators

$$[\Theta^1(\mathbb{CP}^{n+1})], [H_{n+1}], [H_{n+1}]^2, \ldots, [H_{n+1}]^{n+1},$$

thus proving the theorem. $\qquad\square$

Lemma 3.7.8 *For each natural number n, the complex projective space $\mathbb{CP}^n$ imbeds into $\mathbb{CP}^{n+1}$, and the complement is homeomorphic to $\mathbb{R}^{2n+2}$.*

Proof We write elements of complex projective space in homogeneous coordinates (see Example 1.5.10). The inclusion

$$[z_1, z_2, \ldots, z_{n+1}] \mapsto [z_1, z_2, \ldots, z_{n+1}, 0]$$

is an imbedding of $\mathbb{CP}^n$ into $\mathbb{CP}^{n+1}$. The quotient map from $\mathbb{C}^{n+2}$ to $\mathbb{CP}^{n+1}$ restricts to a quotient map $q : S^{2n+3} \longrightarrow \mathbb{CP}^{n+1}$; here we are identifying $\mathbb{C}^{n+2}$ with $\mathbb{R}^{2n+4}$. Consider the set

$$\{(z_1, z_2, \ldots, z_{n+1}, z_{n+2}) \in S^{2n+3} : z_{n+2} \text{ real}\}.$$

This set is homeomorphic to the equator S^{2n+2} of S^{2n+3}, and taking $z_{n+2} > 0$ gives us the northern hemisphere N of S^{2n+2}. The quotient map q sends N into the complement of $\mathbb{CP}^n$ in $\mathbb{CP}^{n+1}$; we show that q is a homeomorphism.

Suppose that $q(z_1, z_2, \ldots, z_{n+2}) = q(w_1, w_2, \ldots, w_{n+2})$ for two points in N. Then from some nonzero complex number λ we see that $z_k = \lambda w_k$ for $1 \leq k \leq n+2$. In particular, $z_{n+2} = \lambda w_{n+2}$. Because both z_{n+2} and w_{n+2} are nonnegative, the number λ must also be nonnegative, and because we are on a unit sphere, the modulus of λ is 1. Therefore $\lambda = 1$, and thus q is injective when restricted to N.

To show surjectivity, take $[z_1, z_2, \ldots, z_{n+2}]$ in $\mathbb{CP}^{n+1} \backslash \mathbb{CP}^n$. Then $z_{n+2} \neq 0$. Write z_{n+2} as $re^{i\theta}$ with $r > 0$. Then

$$
\begin{aligned}
q(e^{-i\theta} z_1, e^{-i\theta} z_2, \ldots, r) &= q(e^{-i\theta}(z_1, z_2, \ldots, z_{n+2})) \\
&= [e^{-i\theta}(z_1, z_2, \ldots, z_{n+2})] \\
&= [z_1, z_2, \ldots, z_{n+2}].
\end{aligned}
$$

Therefore q is a continous bijection, and because it is a quotient map, q is a homeomorphism. The subspace N is homeomorphic to $\mathbb{R}^{2n+2}$, whence the lemma follows. $\qquad\square$

Proposition 3.7.9 *For every compact Hausdorff space X and every nonnegative integer n, external multiplication defines isomorphisms*

$$\mathrm{K}^0(\mathbb{CP}^n) \otimes \mathrm{K}^p(X) \cong \mathrm{K}^0(\mathbb{CP}^n \times X)$$

for $p = 0, -1$.

Proof We proceed by induction on n. The space $\mathbb{CP}^0$ is just a point, in which case the proposition is obviously true. Now suppose the result

holds for n and consider the diagram

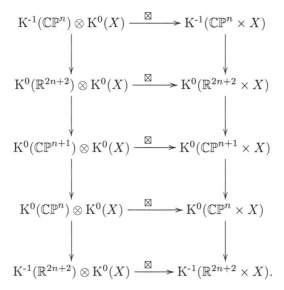

Corollary 2.7.19 and Lemma 3.7.8 imply that the second column is exact. Futhermore, because the groups $K^p(\mathbb{CP}^n)$ are free abelian, Corollary 2.7.19 and Lemma 3.7.8 also imply that the first column is exact. Moreover, direct computation shows that the diagram commutes. The first and fourth horizontal maps are isomorphisms by our inductive hypothesis, and Corollary 3.3.11 implies that the second and fifth horizontal maps are also isomorphisms. Therefore Proposition 1.8.10 gives us

$$K^0(\mathbb{CP}^{n+1}) \otimes K^0(X) \cong K^0(\mathbb{CP}^{n+1} \times X),$$

and induction yields a proof of the desired isomorphism when $p = 0$; a similar argument establishes the isomorphism when $p = -1$. □

Theorem 3.7.10 *Suppose that X is a compact Hausdorff space, let n be a natural number, and suppose that V is a vector bundle of rank n over X. Let H denote the Hopf bundle over $\mathbb{P}(V)$ and let $p : \mathbb{P}(V) \longrightarrow X$ be the projection map. Then $K^0(\mathbb{P}(V)$ is a free $K^0(X)$-module on generators $[\Theta^1(\mathbb{P}(V))], [H], [H]^2, \ldots, [H]^{n-1}$. Moreover, $[H]$ satisfies the single relation*

$$\sum_{i=0}^{n} (-1)^i [\textstyle\bigwedge^i (p^*V)] \otimes [H]^i = 0.$$

In particular, the homomorphism $p^* : K^0(X) \longrightarrow K^0(\mathbb{P}(V))$ is injective.

Proof If V is a trivial bundle, the theorem is a consequence of Propositions 3.7.7 and 3.7.9. The general result then follows from Lemma 3.7.5 and the fact that V is locally trivial.   □

Corollary 3.7.11 (Splitting principle) *Let V be a vector bundle over a compact Hausdorff space X. There exists a compact Hausdorff space $F(V)$ and a continuous function $f : F(V) \longrightarrow X$ such that:*

  (i) *the homomorphism $f^* : \mathrm{K}^0(X) \longrightarrow \mathrm{K}^0(F(V))$ is injective;*
  (ii) *the vector bundle f^*V is isomorphic to the internal Whitney sum of line bundles.*

Proof Without loss of generality, assume that X is connected. We proceed by induction on the rank of V. If V is a line bundle, we take $F(V) = X$ and let f be the identity map. Now suppose that the splitting principle holds for vector bundles of rank less than some natural number n and let V be a vector bundle of rank n. Let $p : \mathbb{P}(V) \longrightarrow X$ be the projection map, and equip $p^*(V)$ with a Hermitian metric. Let $\widetilde{V}$ be the orthogonal complement of H^* in $p^*(V)$. The bundle $\widetilde{V}$ is a rank $n - 1$ vector bundle over $\mathbb{P}(V)$, and we invoke our inductive hypothesis to produce a compact Hausdorff space $F(\widetilde{V})$ and a continuous function $\widetilde{f} : F(\widetilde{V}) \longrightarrow \mathbb{P}(V)$ such that $\widetilde{f}^*$ is injective and such that $\widetilde{f}^*\widetilde{V}$ is isomorphic to the Whitney sum of line bundles. Theorem 3.7.10 states that p^* is an injection, and thus we obtain the splitting principle for V by setting $F(V) = F(\widetilde{V})$ and $f = p\widetilde{f}$.   □

3.8 Operations

For notational convenience in this section, we will often use bold lower case letters to denote elements of K-theory, and we will denote multiplication in K^0 by juxtaposition.

Definition 3.8.1 *An* operation *is a natural transformation from K^0 to itself.*

An operation does not necessarily respect either addition or multiplication on K^0. There are many examples of operations, but we only consider two families of operations in this book. The definition of each family relies on the following fact.

Lemma 3.8.2 *For every commutative ring $\mathcal{R}$ with unit, the set*

$$1 + \mathrm{K}^0[[t]]^+ = \left\{ \sum_{k=0}^{\infty} a_k t^k : a_0 = 1 \right\}$$

is a group under multiplication.

Proof The only nonobvious point to check is the existence of multiplicative inverses. Given a power series $\sum_{k=0}^{\infty} a_k t^k$ in $1 + \mathrm{K}^0[[t]]^+$, we construct its inverse $\sum_{k=0}^{\infty} b_k t^k$.

We proceed by induction on k. Obviously $b_0 = 1$. Suppose we know $b_0, b_1, \ldots, b_k$, multiply the two power series together, and look at the coefficient of t^{k+1}. We have

$$a_{k+1} + a_k b_1 + a_{k-1} b_2 + \cdots + a_1 b_k + b_{k+1} = 0,$$

and thus we can write b_{k+1} in terms of known quantities. $\square$

Construction 3.8.3 *Let X be a compact Hausdorff space, and consider the power series from Definition 3.5.10. For each t, we have a map*

$$\textstyle\bigwedge_t : \mathrm{Vect}(X) \longrightarrow 1 + \mathrm{K}^0[[t]]^+,$$

and Proposition 3.5.11 implies that $\bigwedge_t$ is a monoid homomorphism. Theorem 1.6.7 allows us to extend the domain of $\bigwedge_t$ to $\mathrm{K}^0(X)$. For each positive integer k, the coefficient of t^k determines an operation that is denoted $\bigwedge^k$. By Proposition 3.5.11, we have for each natural number n the equation

$$\textstyle\bigwedge^n (\mathbf{x} + \mathbf{x}') = \sum_{k+l=n} \bigwedge^k (\mathbf{x}) \bigwedge^l (\mathbf{x}')$$

for all $\mathbf{x}$ and $\mathbf{x}'$ in $\mathrm{K}^0(X)$.

Construction 3.8.4 *For each compact Hausdorff space X, define a function $\psi : \mathrm{K}^0(X) \longrightarrow \mathrm{K}^0(X)[[t]]$ by the formula*

$$\psi(\mathbf{x}) = -t \frac{d}{dt} \left(\log \textstyle\bigwedge_{-t}(\mathbf{x}) \right)$$
$$= -t \left(\textstyle\bigwedge_{-t}(\mathbf{x}) \right)^{-1} \frac{d}{dt} \left(\textstyle\bigwedge_{-t}(\mathbf{x}) \right).$$

For each positive integer k, the coefficient of t^k determines an operation that is denoted ψ^k; these are called the Adams operations.

Construction 3.8.3 describes a straightforward way to produce operations, but obviously we must justify defining the ostensibly strange operations in Construction 3.8.4.

Proposition 3.8.5 *Let X be a compact Hausdorff space, let k and l be nonnegative integers, and take $\mathbf{x}$ and $\mathbf{x}'$ in $\mathrm{K}^0(X)$. Then:*

(i) $\psi^k(\mathbf{x} + \mathbf{x}') = \psi^k(\mathbf{x}) + \psi^k(\mathbf{x}')$;

(ii) $\psi^k([L]) = [L]^k$ *for every line bundle L over X;*

(iii) *the Adams operations are uniquely determined by (i) and (ii).*

Proof From Construction 3.8.3 and properties of the logarithm, we have

$$\begin{aligned}
\psi(\mathbf{x} + \mathbf{x}') &= -t\frac{d}{dt}\Big(\log \textstyle\bigwedge_{-t}(\mathbf{x} + \mathbf{x}')\Big) \\
&= -t\frac{d}{dt}\Big(\log\big(\textstyle\bigwedge_{-t}(\mathbf{x})\bigwedge_{-t}(\mathbf{x}')\big)\Big) \\
&= -t\frac{d}{dt}\Big(\log \textstyle\bigwedge_{-t}(\mathbf{x}) + \log \textstyle\bigwedge_{-t}(\mathbf{x}')\Big) \\
&= \psi(\mathbf{x}) + \psi(\mathbf{x}'),
\end{aligned}$$

and this implies (i) for all k.

Observe that $\bigwedge^k(L) = 0$ for all $k > 1$, and therefore

$$\begin{aligned}
\psi[L] &= -t\frac{d}{dt}\log\big(1 - [L]t\big) \\
&= -t\big(1 - [L]t\big)^{-1}\big(-[L]\big) \\
&= [L]t\left(\sum_{k=0}^{\infty}[L]^k t^k\right) \\
&= \sum_{k=1}^{\infty}[L]^k t^k,
\end{aligned}$$

and hence (ii) follows.

To establish (iii), let V be a vector bundle over X. By Corollary 3.7.11, there exists a compact Hausdorff space $F(V)$ and a continuous function $f : F(V) \longrightarrow X$ such that $f^* : \mathrm{K}^0(X) \longrightarrow \mathrm{K}^0(F(V))$ is injective and f^*V is isomorphic to the sum of line bundles. Parts (i) and (ii) therefore determine $\psi[f^*V] = f^*\psi[V]$, and because f^* is injective, they also determine $\psi[V]$. Theorem 1.6.7 then implies that (i) and (ii) determine $\psi(\mathbf{x})$ for all $\mathbf{x}$ in $\mathrm{K}^0(X)$. $\square$

The preceding proposition shows that the Adams operations are well behaved with respect to addition; the next proposition shows that they are also multiplicatively well behaved.

Proposition 3.8.6 *Let X be a compact Hausdorff space, let k, l, and n be nonnegative integers, and take $\mathbf{x}$ and $\mathbf{x}'$ in $\mathrm{K}^0(X)$. Then:*

  (i) $\psi^k(\mathbf{x}\mathbf{x}') = \psi^k(\mathbf{x})\psi^k(\mathbf{x}')$;

  (ii) $\psi^k\psi^l(\mathbf{x}) = \psi^{kl}(\mathbf{x})$;

  (iii) $\psi^2(\mathbf{x}) - \mathbf{x}^2 = 2\mathbf{y}$ *for some* $\mathbf{y}$ *in* $\mathrm{K}^0(X)$;

  (iv) $\psi^k(\mathbf{u}) = k^n\mathbf{u}$ *for every* $\mathbf{u}$ *in* $\widetilde{\mathrm{K}}^0(S^{2n})$.

Proof Theorem 1.6.7, Corollary 3.7.11, and Proposition 3.8.5(iii) imply that we can prove each part of this proposition by verifying it for isomorphism classes of vector bundles that are internal Whitney sums of line bundles. Suppose that $\mathbf{x} = \sum_{i=1}^{m}[L_i]$ and $\mathbf{x}' = \sum_{j=1}^{r}[L_j']$. Tensor products of line bundles are line bundles, so parts (i) and (ii) of Proposition 3.8.5 give us

$$\psi^k(\mathbf{x}\mathbf{x}') = \psi^k\left(\sum_{i,j}[L_i][L_j']\right) = \sum_{i,j}\psi^k\left([L_i][L_j']\right)$$
$$= \sum_{i,j}\psi^k\left([L_i \otimes L_j']\right) = \sum_{i,j}\left([L_i \otimes L_j']\right)^k$$
$$= \sum_{i,j}[L_i]^k[L_j']^k = \psi^k(\mathbf{x})\psi^k(\mathbf{x}')$$

and

$$\psi^k\psi^l(\mathbf{x}) = \psi^k\psi^l\left(\sum_i[L_i]\right) = \psi^k\left(\sum_i\psi^l[L_i]\right)$$
$$= \psi^k\left(\sum_i[L_i]^l\right) = \left(\sum_i\psi^k([L_i]^l)\right) = \sum_i[L_i]^{kl} = \psi^{kl}(\mathbf{x}).$$

We also have

$$\mathbf{x}^2 = \left(\sum_i[L_i]\right)^2 = \sum_{i=1}^{m}[L_i]^2 - 2\sum_{i\neq s}[L_i][L_s] = \psi^2(\mathbf{x}) - 2\sum_{i\neq s}[L_i][L_s],$$

which establishes (iii).

  Consider the element $[H^*] - [1]$ from Example 2.8.2 as an element of

$\widetilde{K}^0(S^2)$. We know from Example 3.3.13 that $([H^*] - [1])^2 = 0$, and thus $([H^*] - [1])^k = 0$ for every integer $k > 1$. Therefore

$$[H^*]^k - 1 = \left(([H^*] - 1) + 1\right)^k - 1 = k([H^*] - 1).$$

Because $[H^*] - [1]$ generates $\widetilde{K}^0(S^2)$, this proves (iv) when $n = 1$. For $n > 1$, identify $K^0(\mathbb{R}^{2n})$ with $\widetilde{K}^0(S^{2n})$; then Corollary 3.3.12, along with the multiplicativity and naturality of the Adams operations, imply the desired result. $\qquad\square$

3.9 The Hopf invariant

We use the results of the previous section to solve a classic problem in topology concerning continuous functions between spheres. We begin by defining two topological spaces that can be associated to any continuous function.

Definition 3.9.1 *Let X and Y be compact Hausdorff spaces and suppose $\phi : X \longrightarrow Y$ is continuous.*

(i) *The* mapping cylinder *of ϕ is the topological space $\mathcal{Z}_\phi$ constructed by identifying $(x, 0)$ in $X \times [0, 1]$ with $\phi(x)$ in Y for all x in X.*

(ii) *The* mapping cone *of ϕ is the pointed topological space $\mathcal{C}_\phi$ constructed by identifying the image of $X \times \{1\}$ in $\mathcal{Z}_\phi$ to a point; this point serves as the basepoint.*

$$\mathcal{Z}_\phi \qquad\qquad\qquad \mathcal{C}_\phi$$

Theorem 3.9.2 (Puppe sequence) *For every continuous function $\phi : X \longrightarrow Y$ between compact Hausdorff spaces, there exists an exact sequence*

$$\widetilde{K}^{-1}(Y) \xrightarrow{\ \phi^*\ } \widetilde{K}^{-1}(X) \longrightarrow \widetilde{K}^0(\mathcal{C}_\phi) \longrightarrow \widetilde{K}^0(Y) \xrightarrow{\ \phi^*\ } \widetilde{K}^0(X).$$

Proof Imbed X into $\mathcal{Z}_\phi$ by $x \mapsto (x, 1)$. From Theorem 2.5.7 we obtain an exact sequence

$$\widetilde{K}^{-1}(\mathcal{Z}_\phi) \longrightarrow \widetilde{K}^{-1}(X) \longrightarrow K^0(\mathcal{Z}_\phi, X) \longrightarrow \widetilde{K}^0(\mathcal{Z}_\phi) \longrightarrow \widetilde{K}^0(X).$$

The map $\psi : \mathcal{Z}_\phi \longrightarrow Y$ that collapses the image of $X \times [0, 1]$ in $\mathcal{Z}_\phi$ to Y is a homotopy inverse to the inclusion of Y into $\mathcal{Z}_\phi$, and therefore $\mathcal{Z}_\phi$ and Y have isomorphic reduced K-theory groups. Our map ϕ is the composition of the inclusion of X into $\mathcal{Z}_\phi$ with the map ψ, and this gives us the homomorphisms ϕ^* at the beginning and end of the sequence in the statement of the theorem. Finally, the space $(\mathcal{Z}_\phi \backslash X)^+$ is homeomorphic to $\mathcal{C}_\phi$, so the remainder of result follows from Theorem 2.2.14 and the definition of reduced K-theory. $\qquad \square$

Suppose that for some natural number n we have a continuous function $\phi : S^{4n-1} \longrightarrow S^{2n}$. Then Theorem 2.6.13, Example 2.8.1, and Theorem 3.9.2 yield a short exact sequence

$$0 \longrightarrow K^0(\mathbb{R}^{4n}) \overset{\alpha}{\longrightarrow} \widetilde{K}^0(\mathcal{C}_\phi) \overset{\gamma}{\longrightarrow} K^0(\mathbb{R}^{2n}) \longrightarrow 0.$$

Corollary 3.3.12 states that $\mathbf{b}^n$ and $\mathbf{b}^{2n}$ are generators of $K^0(\mathbb{R}^{2n})$ and $K^0(\mathbb{R}^{4n})$ respectively, and we can construct a splitting of our short exact sequence by sending $\mathbf{b}^n$ to any element $\mathbf{u}$ in $\widetilde{K}^0(\mathcal{C}_\phi)$ with the property that $\gamma(\mathbf{u}) = \mathbf{b}^n$. Therefore $\widetilde{K}^0(\mathcal{C}_\phi) \cong \mathbb{Z} \oplus \mathbb{Z}$, and if we set $\mathbf{v} = \alpha(\mathbf{b}^{2n})$, then $\mathbf{u}$ and $\mathbf{v}$ are generators of $\widetilde{K}^0(\mathcal{C}_\phi)$. Example 3.3.13 and the fact that S^{2n} is homeomorphic to the one-point compactification of $\mathbb{R}^{2n}$ imply that $(\gamma(\mathbf{u}))^2 = \mathbf{b}^2 = 0$ in $K^0(S^{2n})$ and thus $\mathbf{v} = h(\phi)\mathbf{u}^2$ for some integer $h(\phi)$. This integer is called the *Hopf invariant* of ϕ.

Proposition 3.9.3 *The Hopf invariant is well defined.*

Proof We maintain the notation of the previous paragraph. Suppose that $\gamma(\tilde{\mathbf{u}}) = \mathbf{b}^n$. Then $\tilde{\mathbf{u}} = \mathbf{u} + m\mathbf{v}$ for some integer m. Example 3.3.13 and the injectivity of α imply that $\mathbf{v}^2 = 0$. Moreover, $\gamma(\mathbf{uv}) = \gamma(\mathbf{u})\gamma(\mathbf{v}) = 0$, so $\mathbf{uv}$ is in the kernel of γ. Thus $\mathbf{uv}$ is in the image of α by exactness, and therefore $\mathbf{uv} = k\mathbf{v}$ for some integer k. This implies that $0 = \mathbf{v}^2 = h(\phi)\mathbf{u}^2\mathbf{v} = h(\phi)k\mathbf{uv}$. If $h(\phi) = 0$, then $\mathbf{v} = 0$, contradicting the fact that $\mathbf{v}$ is a generator of $\widetilde{K}^0(\mathcal{C}_\phi)$. Hence $k\mathbf{uv} = 0$, which implies that $\mathbf{uv} = 0$. Therefore

$$\tilde{\mathbf{u}}^2 = (\mathbf{u} + m\mathbf{v})^2 = \mathbf{u}^2 + 2m\mathbf{uv} + m^2\mathbf{v}^2 = \mathbf{u}^2.$$

$\qquad \square$

Theorem 3.9.4 *Suppose that for some natural number n there exists a continuous function $\phi : S^{4n-1} \longrightarrow S^{2n}$ whose Hopf invariant $h(\phi)$ is odd. Then $n = 1, 2,$ or 4.*

Proof By Proposition 3.8.6(iv), we know that $\psi^k(\mathbf{b}^n) = k^n \mathbf{b}^n$ for every positive integer k. The functorality of the Adams operations implies that for each k we have $\psi^k(\mathbf{u}) = k^n \mathbf{u} + m_k \mathbf{v}$ for some integer m_k. Proposition 3.8.6(iii) implies that there exist integers p and q such that

$$\begin{aligned} 2(p\mathbf{u} + q\mathbf{v}) &= \psi^2(\mathbf{u}) - \mathbf{u}^2 \\ &= 2^n \mathbf{u} + m_2 \mathbf{v} - \mathbf{u}^2 \\ &= 2^n \mathbf{u} + m_2 \mathbf{v} - h(\phi)\mathbf{v}. \end{aligned}$$

Comparing the coefficients of $\mathbf{v}$, we see that $m_2 - h(\phi) = 2q$, and because $h(\phi)$ is odd by hypothesis, the integer m_2 is also odd. Next, for every positive integer k, we have

$$\psi^k(\mathbf{v}) = \alpha\big(\psi^k(\mathbf{b}^{2n})\big) = \alpha(k^{2n}\mathbf{b}^{2n}) = k^{2n}\alpha(\mathbf{b}^{2n}) = k^{2n}\mathbf{v}.$$

Proposition 3.8.6(ii) gives us

$$\psi^{2k}(\mathbf{u}) = \psi^k\big(\psi^2(\mathbf{u})\big) = \psi^k(2^n \mathbf{u} + m_2 \mathbf{v}) = 2^n \psi^k(\mathbf{u}) + m_2 \psi^k(\mathbf{v})$$
$$= 2^n(k^n \mathbf{u} + m_k \mathbf{v}) + k^{2n}m_2\mathbf{v} = (2k)^n \mathbf{u} + (2^n m_k + k^{2n}m_2)\mathbf{v},$$

and a similar computation with the roles of 2 and k reversed yields

$$\psi^{2k}(\mathbf{u}) = \psi^2(\psi^k(\mathbf{u})) = (2k)^n \mathbf{u} + (k^n m_2 + 2^{2n}m_k)\mathbf{v}.$$

Thus $2^n m_k + k^{2n}m_2 = k^n m_2 + 2^{2n}m_k$, which implies that $2^n(2^n - 1)m_k = k^n(k^n - 1)m_2$. We know that m_2 is odd, so if $k > 1$ is also odd, the factor 2^n must divide $k^n - 1$. In other words, we seek values of n that solve the congruence $k^n \equiv 1 \mod 2^n$ for all odd values of k.

If $n = 1$, then this congruence holds for all odd k. Now suppose $n > 1$ and consider the multiplicative group $(\mathbb{Z}_{2^n})^*$ of units in the ring of integers modulo 2^n. This group has order 2^{n-1} and thus the order of k in $(\mathbb{Z}_{2^n})^*$ is even by Lagrange's theorem. Lagrange's theorem also implies that the order of k in $(\mathbb{Z}_{2^n})^*$ divides n, whence n is even. Set $k = 1 + 2^{n/2}$. From the binomial theorem we know that $k^n \equiv 1 + n2^{n/2} \mod 2^n$, or, in other words, that 2^n divides $k^n - 1 - n2^{n/2}$. We are assuming that $k^n \equiv 1 \mod 2^n$, so 2^n divides $n2^{n/2}$. But $2^n > n2^{n/2}$ for $n > 5$, and because n is even, we must have $n = 2$ or $n = 4$. $\qquad\square$

3.10 Notes

Most of the results and proofs in this chapter are an amalgam of the presentations in [7] and [12]. Our version of the Künneth theorem is not the most general one possible; see [3].

We have not proved that there are any maps $\phi : S^{4n-1} \longrightarrow S^{2n}$ with odd Hopf invariant when $n = 1, 2, 4$, but, in fact, for each of these values of n, there exists a function with Hopf invariant one. This is closely related to the question of which $\mathbb{R}^n$ admit a reasonable definition of multiplication; the reader should consult [11] for more information.

The original proof of the theorem on the existence of maps with odd Hopf invariant was due to Adams in [1] and did not use K-theory; that paper was over 80 pages long. By constrast, the K-theoretic proof we give is due to Adams and Atiyah ([2]) and is only eight pages long. This was an early example of the power and utility of K-theoretic methods in topology.

There are many other applications of topological K-theory; for some examples, see [14].

Exercises

3.1   For each natural number n, what is $K^{-1}(\mathbb{CP}^n)$?

3.2   Let X be a compact Hausdorff space and suppose that A_1 and A_2 are closed subspaces of X whose union is X and whose intersection is nonempty. Fix a point x_0 in $A_1 \cap A_2$ and take x_0 to be the basepoint of X, A_1, and A_2. Establish a *reduced Mayer–Vietoris* exact sequence

$$\widetilde{K}^0(X) \longrightarrow \widetilde{K}^0(A_1) \oplus \widetilde{K}^0(A_2) \longrightarrow \widetilde{K}^0(A_1 \cap A_2)$$

$$\widetilde{K}^{-1}(A_1 \cap A_2) \longleftarrow \widetilde{K}^{-1}(A_1) \oplus \widetilde{K}^{-1}(A_2) \longleftarrow \widetilde{K}^{-1}(X).$$

3.3   (a) Show that $\mathcal{G} \otimes \mathbb{Z} = 0$ for every finite group $\mathcal{G}$.

        (b) Find a short exact sequence of abelian groups

$$0 \longrightarrow \mathcal{G}_1 \longrightarrow \mathcal{G}_2 \longrightarrow \mathcal{G}_3 \longrightarrow 0$$

with the property that the sequence

$$0 \longrightarrow \mathcal{G}_1 \otimes \mathbb{Z} \longrightarrow \mathcal{G}_2 \otimes \mathbb{Z} \longrightarrow \mathcal{G}_3 \otimes \mathbb{Z} \longrightarrow 0$$

is not exact.

3.4 Suppose that X is a compact Hausdorff space and that V and W are vector bundles over X. Prove that $\text{Hom}(V, W) \cong V^* \otimes W$ (see Exercise 2.4).

3.5 Let X be a connected compact Hausdorff space with basepoint x_0, and suppose that X is a union of n contractible closed sets. Prove that the product of any n elements in $\widetilde{K}^0(X, x_0)$ is zero.

3.6 Let X and Y be nonempty compact Hausdorff spaces. The *join* of $X*Y$ is the topological space obtained by forming the product $X \times Y \times [0, 1]$ and making the identifications:

- $(x, y, 0) \sim (x', y', 0)$ if $y = y'$;
- $(x, y, 1) \sim (x', y', 1)$ if $x = x'$.

 (a) For each x in X, define $i_X(x)$ to be the image of $(x, y, 0)$ in $X * Y$ for any y. Show that i_X is well defined and is am imbedding of X into $X * Y$. Similarly, for each y in Y, define $i_Y(y)$ to be the image of $(x, y, 1)$ in $X * Y$, and show that i_Y is a well defined imbedding.

 (b) Prove that i_X and i_Y are homotopic to constant functions.

 (c) Prove that for $p = 0, -1$, there is a short exact sequence

$$0 \longrightarrow \widetilde{K}^{-(p+1)}(X * Y) \longrightarrow \widetilde{K}^p(X \times Y)$$

$$\longrightarrow \widetilde{K}^p(X \times Y) \longrightarrow \widetilde{K}^p(X) \oplus \widetilde{K}^p(Y) \longrightarrow 0.$$

 (d) Suppose that X of finite type and that $K^0(Y)$ and $K^{-1}(Y)$ are free abelian groups. Prove that there exist isomorphisms

$$\widetilde{K}^0(X * Y) \cong (\widetilde{K}^0(X) \otimes \widetilde{K}^{-1}(Y)) \oplus (\widetilde{K}^0(X) \otimes \widetilde{K}^{-1}(Y))$$

$$\widetilde{K}^{-1}(X * Y) \cong (\widetilde{K}^0(X) \otimes \widetilde{K}^0(Y)) \oplus (\widetilde{K}^{-1}(X) \otimes \widetilde{K}^{-1}(Y)).$$

3.7 A finitely generated $\mathcal{R}$-module $\mathcal{P}$ is *projective* if it satisfies the following property: given a surjective $\mathcal{R}$-module homomorphism $\phi : \widetilde{\mathcal{M}} \longrightarrow \mathcal{M}$ and any $\mathcal{R}$-module homomorphism $\psi : \mathcal{P} \longrightarrow \mathcal{M}$, there exists an $\mathcal{R}$-module homomorphism $\widetilde{\psi} : \mathcal{P} \longrightarrow \widetilde{\mathcal{M}}$ such that $\psi = \phi\widetilde{\psi}$.

 (a) Prove that a free $\mathcal{R}$-module of rank n is projective for every natural number n.

 (b) Prove that $\mathcal{P}$ is projective if and only if there exists a $\mathcal{R}$-module $\mathcal{Q}$ such that $\mathcal{P} \oplus \mathcal{Q}$ is free.

3.8 Let V be a vector bundle over a compact Hausdorff space X, and let $\text{Sect}(V)$ denote the collection of sections of V.

    (a) Show that $\text{Sect}(V)$ is a $C(X)$-module, and that $\text{Sect}(V)$ is a finitely generated free module if and only if V is trivial.

    (b) Prove that $\text{Sect}(V)$ is finitely generated and projective.

    (c) Verify that the correspondence $V \mapsto \text{Sect}(V)$ determines a covariant functor from the category of vector bundles over X and bundle homomorphisms to the category of finitely generated projective $C(X)$-modules and module homomorphisms.

    (d) Let $\mathcal{P}$ be a finitely generated projective C(X)-module. Show that $\mathcal{P} \cong \text{Sect}(V)$ for some vector bundle V over X.

    (e) (Serre–Swan theorem) Let $\text{Proj}(C(X))$ denote the collection of isomorphism classes of finitely generated projective modules over X. Verify that direct sum makes $\text{Proj}(X)$ into an abelian monoid, and show that there is a monoid isomorphism from $\text{Vect}(X)$ to $\text{Proj}(C(X))$.

3.9 Suppose X and Y are compact Hausdorff spaces. Use Theorem 2.6.13 to identify $\text{K}^{-1}(Y)$ and $\text{K}^{-1}(X \times Y)$ with $\text{K}^0(Y \times \mathbb{R})$ and $\text{K}^0(X \times Y \times \mathbb{R})$ respectively. Show that the external product

$$\text{K}^0(X) \times \text{K}^{-1}(Y) \longrightarrow \text{K}^{-1}(X \times Y)$$

has the following formula: take an idempotent E in $\text{M}(m, C(X))$ and an element S in $\text{GL}(n, C(Y))$. Then

$$[\mathsf{E}] \boxtimes [\mathsf{S}] = \big[(\mathsf{E} \boxtimes I_n)(I_m \boxtimes \mathsf{S})(\mathsf{E} \boxtimes I_n) + (I_m - \mathsf{E}) \boxtimes I_n\big].$$

3.10 Let X be a compact Hausdorff space, take $\mathbf{x}$ in $\text{K}^0(X)$, and let p be a prime number. Prove that

$$\psi^p(\mathbf{x}) \equiv \mathbf{x}^p \qquad \mod p.$$

In other words, show that $\psi^p(\mathbf{x}) - \mathbf{x}^p = p\widetilde{\mathbf{x}}$ for some $\widetilde{\mathbf{x}}$ in $\text{K}^0(X)$.

4

Characteristic classes

In general, it is quite difficult to tell if two vector bundles are isomorphic. The theory of *characteristic classes* is designed to remedy this problem. The idea of characteristic classes is to assign to each isomorphism class of vector bundles (or more generally each K-theory class) an element in a cohomology theory in which it is easier to distinguish distinct objects.

While we could define characteristic classes quite generally, we will simplify matters greatly and only consider *smooth* vector bundles over *smooth* compact manifolds. Because not all readers may have had a class in differential topology, we begin this chapter with a crash course in the subject, paying particular attention to the cohomology theory we will use, *de Rham cohomology*.

4.1 De Rham cohomology

Let $x_1, x_2, \ldots, x_n$ be the standard coordinate functions on $\mathbb{R}^n$ and let $U \subseteq \mathbb{R}^n$ be an open set. A real-valued function on U is *smooth* if it possesses partial derivatives of all orders. A complex-valued function is smooth if its real and imaginary parts are both smooth; the reader is warned that this is not the same as saying a function is holomorphic.

For each of the coordinates x_k, associate an object dx_k, and let $\Omega^1(U)$ denote the set of formal sums

$$\Omega^1(U) = \left\{ \sum_{k=1}^n f_k \, dx_k : \text{each } f_k \text{ a smooth } \mathbb{C}\text{-valued function on } U \right\}.$$

The elements of $\Omega^1(U)$ are called *(complex) differential one-forms*, or usually just *one-forms*. One-forms are added in the obvious way, which makes $\Omega^1(U)$ into a vector space. We then define $\Omega^0(U)$ to be the collection of smooth functions on U and set $\Omega^m(U) = \bigwedge^m(\Omega^1(U))$ for each natural number m. Elements of $\Omega^m(U)$ are called *m-forms* or *forms of degree m*. Finally, we define $\Omega^*(U)$ to be $\bigwedge(\Omega^1(U))$, and we usually denote the wedge of forms by juxtaposition.

There are two important operations on $\Omega^*(U)$. The first of these is *integration*. In vector calculus, integration is defined by taking limits of Riemann sums; we denote this integral by $\int_U f\,|dx_1\,dx_2\,\ldots\,dx_n|$ to highlight the fact that in the Riemann integral, the order of the dx_k does not matter. However, the order does matter in the form $f\,dx_1\,dx_2\,\ldots\,dx_n$. For this reason, given a permutation ρ of $\{1,2,\ldots,n\}$ we define the integral of the n-form $f\,dx_1\,dx_2\,\ldots\,dx_n$ so that it satisfies the equation

$$\int_U f\,dx_{\rho(1)}\,dx_{\rho(2)}\,\ldots\,dx_{\rho(n)} = \operatorname{sign}\rho \int_U f\,|dx_1\,dx_2\,\ldots\,dx_n|,$$

where $\operatorname{sign}\rho$ is 1 if ρ is an even permuation and -1 if it is an odd permutation.

The other operation on $\Omega^*(U)$ we consider is *exterior differentiation*, denoted d.

For zero-forms (smooth functions) f, the exterior derivative is defined by the formula

$$df = \sum_{k=1}^n \frac{\partial f}{\partial x_k} dx_k.$$

More generally, given an m-form $\omega = f\,dx_{k_1}\,dx_{k_2}\,\ldots\,dx_{k_m}$, its exterior derivative is the $(m+1)$-form $d\omega = df\,dx_{k_1}\,dx_{k_2}\,\ldots\,dx_{k_m}$; we extend the definition of d to general m-forms by linearity.

There are three crucial properties of the exterior derivative. First, d is an *antiderivation*, which means that given a k-form α and an l-form β, we have

$$d(\alpha\,\beta) = (d\alpha)\,\beta + (-1)^k \alpha\,(d\beta).$$

The second important property of d is that

$$d^2(\omega) = d(d\omega) = 0$$

for every $\omega \in \Omega^*(U)$; this follows from the fact that d is an antiderivation and that mixed partial derivatives are equal.

Finally, suppose we have open sets $U \subseteq \mathbb{R}^n$ and $V \subseteq \mathbb{R}^p$ and a smooth map $\psi : U \longrightarrow V$ (in other words, each component of ψ is a smooth real-valued function). Then given a smooth function f on V, we can define the *pullback* of f to be the smooth function $\psi^* f$ on U given by the formula

$$(\psi^* f)(x) = f(\psi(x)).$$

To extend the notion of pullback to forms, take the standard coordinate functions $x_1, x_2, \ldots, x_n$ and $y_1, y_2, \ldots, y_p$ on U and V respectively. Then for each m, define $\psi^* : \Omega^m(V) \longrightarrow \Omega^m(U)$ by requiring that ψ^* be linear and that

$$\psi^* (f \, dy_{k_1} \, dy_{k_2} \, \ldots \, dy_{k_m}) = (\psi^* f) \, d(\psi^* y_{k_1}) \, d(\psi^* y_{k_2}) \, \ldots \, d(\psi^* y_{k_m}). \quad (\star)$$

A computation using the chain rule shows that $d(\psi^* \omega) = \psi^*(d\omega)$ for all ω in $\Omega^*(V)$; this fact is often stated "the exterior derivative commutes with pullbacks."

Our next goal is to extend the notion of differential forms to spaces more general than open subsets of $\mathbb{R}^n$. Let M be a topological space that is Hausdorff and second countable. A *chart* on M is a pair (U, ϕ), where U is an open subset of M and ϕ is an imbedding of U into $\mathbb{R}^n$ for some n. Two charts (U_1, ϕ_1) and (U_2, ϕ_2) on M are *smoothly related* if the composition $\phi_1 \circ \phi_2^{-1} : \phi_2(U_2) \longrightarrow \phi_1(U)$ is a smooth map. An *atlas* for M is a collection $\mathcal{A}$ of smoothly related charts that cover M, and a *maximal atlas* is an atlas $\mathcal{A}$ that contains every chart that is smoothly related to some chart (and therefore every chart) in $\mathcal{A}$; any atlas can be enlarged to a maximal one. If M admits a maximal atlas $\mathcal{A}$, the pair $(M, \mathcal{A})$ is called a *smooth manifold*. We usually suppress the atlas and the word "smooth" and simply refer to M as a manifold. If all the charts of M map into the same $\mathbb{R}^n$ (which is necessarily the case if M is connected), we say M is *n-dimensional*. In this chapter, we will assume that our manifolds are compact.

A complex-valued function f on a manifold M is *smooth* if for each chart (U, ϕ) in the atlas, the composition $(f|_U) \circ \phi^{-1} : \phi(U) \longrightarrow \mathbb{C}$ is smooth.

To define forms on M, we begin by defining them on charts. Let (U, ϕ) be a chart on M, let $x_1, x_2, \ldots, x_n$ be the standard coordinate functions on $\mathbb{R}^n$ restricted to $\phi(U)$, and for each $1 \leq k \leq n$, let $y_k = x_k \circ \phi$. The functions y_k are called *local coordinates*, and an *m-form* on U is a linear combination of terms that look like $f \, dy_{k_1} \, dy_{k_2} \ldots dy_{k_m}$ for some smooth function f on U. For each f and each point y in U, we define

$$\frac{\partial f}{\partial y_k}(y) = \frac{\partial \left(f \circ \phi^{-1} \right)}{\partial x_k}(\phi(y)).$$

Then

$$df = \sum_{k=1}^{n} \frac{\partial f}{\partial y_k} dy_k,$$

and for every m-form $\omega = f\, dy_{k_1}\, dy_{k_2} \ldots dy_{k_m}$, we have

$$d\omega = df\, dy_{k_1}\, dy_{k_2} \ldots dy_{k_m}.$$

Let (U, ϕ_U) and (V, ϕ_V) be charts on manifolds M and N, respectively, let ω be a differential form on V, and suppose $\psi : U \longrightarrow V$ is smooth. Then the pullback $\psi^* f \in \Omega^0(U)$ of a smooth function $f \in \Omega^0(V)$ still makes sense, and extends as in $(\star)$ to allow us to pull back forms. If $\mathcal{A}$ denotes the maximal atlas for M, we define a form ω on M to be a collection of forms $\{\omega_U \in \Omega^*(U) : U \in \mathcal{A}\}$ with the following property: for every pair (U_1, ϕ_1) and (U_2, ϕ_2) of charts on M such that $U_1 \cap U_2$ is nonempty, define i_1 and i_2 to be the inclusions of $U_1 \cap U_2$ into U_1 and U_2 respectively. Then $i_1^*(\omega_{U_1}) = i_2^*(\omega_{U_2})$.

The notions of pullback and exterior derivative carry over to the set $\Omega^*(M)$ of forms on M, and it is still true that $d^2 = 0$ and that the exterior derivative commutes with pullbacks.

We can also integrate forms on manifolds. The first step is to see how the integral transforms under change of variables. Let U and $\widetilde{U}$ be open subsets of $\mathbb{R}^n$, and let $\phi : U \longrightarrow \widetilde{U}$ be a smooth map with an inverse that is also smooth. Then a computation using the change of variables formula from vector calculus shows that

$$\int_U \phi^* \widetilde{\omega} = \pm \int_{\widetilde{U}} \widetilde{\omega}$$

for every n-form $\widetilde{\omega}$ on $\widetilde{U}$. If the integrals above are equal, we say that ϕ is *orientation-preserving*, and if the integrals differ by a sign, we say ϕ is *orientation-reversing*.

In order to define integration on a manifold M, we must be able to choose an open cover of M consisting of charts (U_α, ϕ_α) with the feature that whenever $U_\alpha \cap U_\beta$ is nonempty, the composition $\phi_\alpha \circ \phi_\beta^{-1}$ is orientation-preserving. This is not always possible, but if it is, we say that M is *orientable*. Many manifolds are orientable, but the real projective plane, for example, is one that is not. Even if M is orientable, the orientation is not unique; when M is connected, there are two distinct orientations. Roughly speaking, an orientation amounts to consistent ordering of the dy_i in the local coordinates of M. For the manifold $\mathbb{R}^n$, one orientation corresponds to the n-form $dx_1\, dx_2 \ldots dx_n$ (or more generally, $dx_{\rho(1)}\, dx_{\rho(2)} \ldots dx_{\rho(n)}$ for any even permutation ρ), and the other orientation corresponds to the n-form $dx_2\, dx_1 \ldots dx_n$ (or more generally, $dx_{\rho(1)}\, dx_{\rho(2)} \ldots dx_{\rho(n)}$ for any odd permutation ρ). If we have specified an orientation for M, we say M is *oriented*.

Suppose M is oriented. Choose a collection of orientation-preserving charts $\mathcal{U} = \{U_\alpha, \phi_\alpha\}$ that cover M and choose a partition of unity $\{\rho_\alpha\}$ subordinate to $\mathcal{U}$ such that each ρ_α is smooth; such partitions of unity always exist. Then for any n-form ω, we define

$$\int_M \omega = \sum_\alpha \left(\int_{\phi_\alpha(U_\alpha)} \left(\phi_\alpha^{-1}\right)^* (\rho_\alpha \omega) \right).$$

The integral is independent of the choice of $\mathcal{U}$ and the partition of unity $\{\rho_\alpha\}$. If the orientation on M is reversed, this integral changes by a sign.

If ω is a form that has degree not equal to n, we declare the integral of ω to be 0.

The most important theorem concerning integration of forms on manifolds is the following.

Theorem 4.1.1 (Stokes' theorem) *Let ω be a differential form on a manifold M. Then*

$$\int_M d\omega = 0.$$

We know what differential forms look like locally, but what about their global behavior? An important tool in understanding the large-scale structure of forms is *de Rham cohomology*, which we now discuss.

Definition 4.1.2 *Let M be a manifold. A differential form ω on M is closed if $d\omega = 0$ and exact if $\omega = d\alpha$ for some form α on M. The set of closed m-forms on M is denoted $Z^m(M)$, and the set of exact m-forms on M is denoted $B^m(M)$.*

Both $Z^m(M)$ and $B^m(M)$ are vector spaces for every nonnegative integer m. Moreover, because $d^2 = 0$, every exact form is closed, and so $B^m(M)$ is a vector subspace of $Z^m(M)$. The elements of $B^m(M)$ are viewed as being "trivially" closed. The search for more interesting closed forms prompts the following definition.

Definition 4.1.3 (de Rham cohomology) *Let M be a compact manifold. For each natural number m, the mth de Rham cohomology group of M is*

$$H^m_{deR}(M) = \frac{Z^m(M)}{B^m(M)}.$$

The direct sum of the de Rham cohomology groups of M is denoted
$H_{deR}^*(M)$.

Proposition 4.1.4 *For every compact manifold M, the wedge product
defines a ring structure on $H_{deR}^*(M)$.*

Proof We show that the wedge of an exact form with a closed form is
exact; the proposition easily follows from this fact. Suppose that ω is
exact and that β is closed. Choose a form α such that $\omega = d\alpha$. Then

$$d(\alpha\beta) = (d\alpha)\beta \pm \alpha d(\beta) = (d\alpha)\beta = \omega\beta.$$

$\square$

Our interest in de Rham cohomology is as a tool to distinguish K-
theory classes, so we will not discuss this theory in any great detail.
However, here is a list of some of the important features of de Rham
theory:

- If M is a manifold of dimension n, then $H_{deR}^m(M) = 0$ for $m > n$.
- If M is connected and has dimension n, then $H_{deR}^n(M) = \mathbb{C}$ if M is
  orientable and 0 if M is nonorientable; in the former case, a representative of $H_{deR}^n(M)$ is called a *volume form*.
- Let $\psi : M \longrightarrow N$ be a smooth map between manifolds. Then for each
  nonnegative integer k, the pullback map ψ^* from $\Omega^k(N)$ to $\Omega^k(M)$
  induces a group homomorphism $\psi^* : H_{deR}^k(N) \longrightarrow H_{deR}^k(M)$.
- If M and N are homotopy equivalent, then the groups $H_{deR}^*(M)$ and
  $H_{deR}^*(N)$ are isomorphic.

4.2 Invariant polynomials

In this section we gather some definitions and facts about polynomials
that we will need in the next section.

Definition 4.2.1 *Let n be a natural number. A function*

$$P : \mathrm{M}(n, \mathbb{C}) \longrightarrow \mathbb{C}$$

is called an invariant polynomial *if P is a polynomial function of the
entries of the matrix and $P(BAB^{-1}) = P(A)$ for all A in $\mathrm{M}(n, \mathbb{C})$ and
B in $\mathrm{GL}(n, \mathbb{C})$. The set of invariant polynomials on $\mathrm{M}(n, \mathbb{C})$ is denoted
$\mathcal{IP}(n, \mathbb{C})$.*

The constant functions are obviously invariant polynomials. The following construction provides some more interesting examples.

Proposition 4.2.2 *Let n be a natural number, and for each A in $M(n, \mathbb{C})$, define $c_1(\mathsf{A})$, $c_2(\mathsf{A})$, $\ldots$, $c_n(\mathsf{A})$ by the equation*

$$\det (1 + x\mathsf{A}) = 1 + c_1(\mathsf{A})x + c_2(\mathsf{A})x^2 + \cdots + c_m(\mathsf{A})x^n.$$

Then for each $1 \leq k \leq n$, the function c_k is an invariant polynomial.

Proof An inspection of the equation in the statement of the proposition shows that each $c_k(\mathsf{A})$ is a polynomial in the entries of A, and the similarity invariance of the determinant implies that each c_k is an invariant polynomial. $\square$

Observe that $c_1(\mathsf{A})$ and $c_n(\mathsf{A})$ are the trace and determinant of A respectively.

Proposition 4.2.3 *An invariant polynomial is determined by its values on diagonal matrices.*

Proof Any invariant polynomial $P : M(n, \mathbb{C}) \longrightarrow \mathbb{C}$ is a continuous function. Therefore to prove the proposition, we need only show that every element of $M(n, \mathbb{C})$ is a limit of matrices with distinct eigenvalues, because such matrices are similar to diagonal matrices.

Let A be in $M(n, \mathbb{C})$. Then A is similar to an upper triangular matrix

$$\mathsf{T} = \begin{pmatrix} t_{11} & t_{12} & t_{13} & \cdots & t_{1n} \\ 0 & t_{22} & t_{23} & \cdots & t_{2n} \\ 0 & 0 & t_{33} & \cdots & t_{3n} \\ \vdots & \vdots & \vdots & \vdots & \vdots \\ 0 & 0 & 0 & \cdots & t_{nn} \end{pmatrix},$$

and so the eigenvalues of T, and therefore A, are $t_{11}, t_{22}, \ldots, t_{nn}$. For all sufficiently small $\epsilon > 0$, the matrix

$$\mathsf{T}_\epsilon = \begin{pmatrix} t_{11} + \epsilon & t_{12} & t_{13} & \cdots & t_{1n} \\ 0 & t_{22} + 2\epsilon & t_{23} & \cdots & t_{2n} \\ 0 & 0 & t_{33} + 3\epsilon & \cdots & t_{3n} \\ \vdots & \vdots & \vdots & \vdots & \vdots \\ 0 & 0 & 0 & \cdots & t_{nn} + n\epsilon \end{pmatrix}$$

has distinct eigenvalues. $\square$

Definition 4.2.4 *Let n be a natural number and let $\mathbb{C}[x_1, x_2, \ldots, x_n]$ be the ring of complex polynomials in the indeterminates $x_1, x_2, \ldots, x_n$. A (complex) symmetric function is a polynomial μ in $\mathbb{C}[x_1, x_2, \ldots, x_n]$ such that*

$$\mu(x_{\rho(1)}, x_{\rho(2)}, \ldots, x_{\rho(n)}) = \mu(x_1, x_2, \ldots, x_n)$$

for every permutation ρ of the set $\{1, 2, \ldots, n\}$.

Our interest in symmetric functions stems from the next result.

Lemma 4.2.5 *Let n be a natural number and suppose that P is an invariant polynomial on $\mathrm{M}(n, \mathbb{C})$. Then there exists a symmetric function μ_P in $\mathbb{C}[x_1, x_2, \ldots, x_n]$ such that*

$$P(\mathrm{diag}(d_1, d_2, \ldots, d_n)) = \mu_P(d_1, d_2, \ldots, d_n)$$

for every diagonal matrix $\mathrm{diag}(d_1, d_2, \ldots, d_n)$.

Proof For every permutation ρ of the set $\{1, 2, \ldots, n\}$, Lemma 1.7.1 states that there is a matrix S in $\mathrm{GL}(n, \mathbb{C})$ such that

$$\mathsf{S}\,\mathrm{diag}(d_1, d_2, \ldots, d_n)\mathsf{S}^{-1} = \mathrm{diag}(d_{\rho(1)}, d_{\rho(2)}, \ldots, d_{\rho(n)})$$

for every diagonal matrix. $\qquad\square$

In light of Lemma 4.2.5, we seek more information about symmetric functions.

Definition 4.2.6 *Let n be a natural number. For $1 \le k \le n$, define*

$$R_k = \{(i_1, i_2, \ldots, i_k) : 1 \le i_1 < i_2 < \cdots < i_k \le n\}.$$

The kth elementary symmetric function in $\mathbb{C}[x_1, x_2, \ldots, x_n]$ is defined by the formula

$$\sigma_k(x_1, x_2, \ldots, x_n) = \sum_{R_k} x_{i_1} x_{i_2} \cdots x_{i_k}.$$

For example, for $n = 4$, the elementary symmetric functions are

$$\sigma_1 = x_1 + x_2 + x_3 + x_4$$
$$\sigma_2 = x_1 x_2 + x_1 x_3 + x_1 x_4 + x_2 x_3 + x_2 x_4 + x_3 x_4$$
$$\sigma_3 = x_1 x_2 x_3 + x_1 x_2 x_4 + x_1 x_3 x_4 + x_2 x_3 x_4$$
$$\sigma_4 = x_1 x_2 x_3 x_4.$$

Note that each elementary symmetric function is, in fact, symmetric.

Proposition 4.2.7 *Let n be a natural number. Then for all $1 \leq k \leq n$, we have*

$$c_k(\mathrm{diag}(d_1, d_2, \ldots, d_n)) = \sigma_k(d_1, d_2, \ldots, d_n)$$

for every diagonal matrix $\mathrm{diag}(d_1, d_2, \ldots, d_n)$.

Proof Apply Proposition 4.2.2 to $\mathrm{diag}(d_1, d_2, \ldots, d_n)$, multiply out the left side, and collect terms. $\qquad\square$

Proposition 4.2.8 *Every symmetric function is a complex polynomial in the elementary symmetric functions.*

Proof Fix n, let $p_1, p_2, \ldots, p_n$ be the first n prime numbers in ascending order, and for each monomial $x_1^{a_1} x_2^{a_2} \cdots x_n^{a_n}$, set

$$\phi(x_1^{a_1} x_2^{a_2} \cdots x_n^{a_n}) = p_1^{a_1} p_2^{a_2} \cdots p_n^{a_n}.$$

Given any symmetric function μ in $\mathbb{C}[x_1, x_2, \ldots, x_n]$, let $\Phi(\mu)$ be the maximum value of ϕ applied to every (nonzero) monomial of μ. In particular, Φ does not detect the coefficients of the monomials. We prove the proposition by induction on the value of Φ.

Suppose $\Phi(\mu) = 1$. Then μ is necessarily a constant monomial, and therefore is trivially a complex polynomial in the elementary symmetric functions. Now suppose that the proposition is true for all symmetric functions μ such that $\Phi(\mu) < N$, and choose a symmetric function ν such that $\Phi(\nu) = N = p_1^{s_1} p_2^{s_2} \cdots p_n^{s_n}$; thus the "largest" monomial in ν is $x_1^{s_1} x_2^{s_2} \cdots x_n^{s_n}$.

For each elementary symmetric function σ_k, we have

$$\Phi(\sigma_k) = \phi(x_{n-k+1} x_{n-k+2} \cdots x_{n-1} x_n) = p_{n-k+1} p_{n-k+2} \cdots p_{n-1} p_n.$$

Moreover, the function Φ is multiplicative on the elementary symmetric functions, and therefore

$$\Phi(\sigma_1^{t_1} \sigma_2^{t_2} \cdots \sigma_n^{t_n}) = p_1^{t_n} p_2^{t_n+t_{n-1}} \cdots p_{n-1}^{t_n+t_{n-1}+\cdots+t_2} p_n^{t_n+t_{n-1}+\cdots+t_2+t_1}$$

for all natural numbers $t_1, t_2, \ldots, t_n$. Comparing this expression with the factorization of N, we see that

$$\Phi\left(\sigma_1^{s_n-s_{n-1}} \sigma_2^{s_{n-1}-s_{n-2}} \cdots \sigma_{n-1}^{s_2-s_1} \sigma_n^{s_1}\right) = N.$$

Let α be the coefficient of the monomial $x_1^{s_1} x_2^{s_2} \cdots x_n^{s_n}$ in ν. Then

$$\nu - \alpha \sigma_1^{s_n-s_{n-1}} \sigma_2^{s_{n-1}-s_{n-2}} \cdots \sigma_{n-1}^{s_2-s_1} \sigma_n^{s_1}$$

is a symmetric function, and

$$\Phi\left(\nu - \alpha\sigma_1^{s_n - s_{n-1}}\sigma_2^{s_{n-1}-s_{n-2}}\cdots\sigma_{n-1}^{s_2-s_1}\sigma_n^{s_1}\right) < N,$$

whence the result follows. □

Theorem 4.2.9 *For each natural number n, the set $\mathcal{IP}(n,\mathbb{C})$ is a ring generated by $c_1, c_2, \ldots, c_n$.*

Proof It is easy to see that the set of invariant polynomials forms a ring. The rest of the theorem follows from Proposition 4.2.3, Lemma 4.2.5, Proposition 4.2.7, and Proposition 4.2.8. □

Proposition 4.2.10 (Newton's identities) *Let n be a natural number and define*

$$b_k = x_1^k + x_2^k + \cdots + x_n^k$$
$$c_k = \sigma_k(x_1, x_2, \ldots, x_n)$$

for $1 \le k \le n$. Then

$$b_k + \sum_{j=1}^{k-1}(-1)^j c_j b_{k-j} + (-1)^k k c_k = 0 \qquad (*)$$

for each k.

Proof For notational convenience in this proof, write each b_k and c_k as $b_{k,n}$ and $c_{k,n}$ respectively and let $F_{k,n}$ denote the left-hand side of $(*)$. We prove the theorem by induction on $n - k$.

Suppose that $k = n$, and define

$$f(x) = \prod_{j=1}^{n}(x - x_j)$$

$$= x^n + \sum_{j=1}^{n-1} c_{j,n} x^{n-j} + (-1)^n c_{n,n},$$

where the second equality is obtained by expanding f. We have

$$0 = f(x_1) = x_1^n + \sum_{j=1}^{n-1}(-1)^j c_{j,n} x_1^{n-j} + (-1)^n c_{n,n}$$

$$0 = f(x_2) = x_2^n + \sum_{j=1}^{n-1}(-1)^j c_{j,n} x_2^{n-j} + (-1)^n c_{n,n}$$

$$\vdots$$

$$0 = f(x_n) = x_n^n + \sum_{j=1}^{n-1}(-1)^j c_{j,n} x_n^{n-j} + (-1)^n c_{n,n}.$$

Adding together these n equations gives us

$$0 = b_{n,n} + \sum_{j=1}^{n-1}(-1)^j c_{j,n} b_{n-j,n} + (-1)^n n c_{n,n},$$

and so the result holds for $n - k = 0$.

Now suppose the theorem is true for $n - k < N \le n$, and consider the case $n - k = N$. Observe that

$$b_{j,n}(x_1, x_2, \ldots, x_{n-1}, 0) = b_{j,n-1}(x_1, x_2, \ldots, x_{n-1})$$
$$c_{j,n}(x_1, x_2, \ldots, x_{n-1}, 0) = c_{j,n-1}(x_1, x_2, \ldots, x_{n-1})$$

for all $1 \le j \le n$. Thus

$$F_{k,n}(x_1, x_2, \ldots, x_{n-1}, 0) = F_{k,n-1}(x_1, x_2, \ldots, x_{n-1}). \qquad (**)$$

Because $n - 1 - k = N - 1 < N$, our inductive hypothesis implies that right-hand side of $(**)$ is zero. This implies that x_n is a factor of $F_{k,n}$. Because $F_{k,n}$ is a symmetric polynomial, the monomials $x_1, x_2, \ldots, x_{n-1}$ are also factors of $F_{k,n}$. Therefore $F_{k,n}$ must be identically zero, because otherwise $F_{k,n}$ would be a polynomial of degree $k < n$ with n distinct factors. $\qquad\square$

Corollary 4.2.11 *Let n be a natural number. For each $1 \le k \le n$ and each matrix A in $M(n, \mathbb{C})$, define $b_k(A) = \operatorname{Tr}(A^k)$. Then each b_k is an invariant polynomial on $M(n, \mathbb{C})$, and*

$$b_k(A) + \sum_{j=1}^{k-1}(-1)^j c_j(A) b_{k-j}(A) + (-1)^k k c_k(A) = 0.$$

Proof The similarity invariance of the trace immediately implies that each b_k is an invariant polynomial. Proposition 4.2.10 tells us that the equation above holds for diagonal matrices, and Proposition 4.2.3 yields the desired result for general matrices. □

Corollary 4.2.12 *For each natural number n, the set $\mathcal{IP}(n, \mathbb{C})$ is a ring generated by $b_1, b_2, \ldots, b_n$.*

Proof For each n, Proposition 4.2.10 implies that the set $\{b_1, b_2, \ldots, b_n\}$ determines the set $\{c_1, c_2, \ldots, c_n\}$, and conversely; the corollary is therefore a consequence of Theorem 4.2.9. □

4.3 Characteristic classes

Given two vector bundles over the same topological space, how can we tell if they are isomorphic or not? If the vector bundles have different ranks, then they are not isomorphic, but at this point in the book, we have no other way of distinguishing vector bundles. In this section we define and discuss the theory of characteristic classes, which is an important tool for answering this question. The easiest approach to this theory is generally known as *Chern–Weil theory*, and involves differentiation. For this reason, we only consider topological spaces M that are manifolds, and we will assume that our vector bundles are smooth. *A priori*, it is not obvious that every element of $K^0(M)$ can be represented by a difference of smooth vector bundles. Fortunately, the additional hypothesis of smoothness does not affect K-theory. More precisely, the following statements are true:

- Every vector bundle over a compact manifold admits a unique smooth structure.
- If two vector bundles are isomorphic via a continuous isomorphism, they are isomorphic via a smooth isomorphism.
- Every idempotent over a compact manifold is similar to a smooth idempotent.
- If two smooth idempotents are homotopic via a continuous homotopy of idempotents, then they are homotopic via a smooth homotopy of idempotents.

The fact that continuous functions can be approximated by smooth ones makes these statements plausible, but the proofs are not particularly edifying and would take us somewhat far afield, so we simply state the results without proof.

Definition 4.3.1 *Let M be a compact manifold and let n be a natural number. For each $\alpha = (\alpha_1, \alpha_2, \ldots, \alpha_n)$ in $(\Omega^*(M))^n$, define the function* $\mathsf{D} : (\Omega^*(M))^n \longrightarrow (\Omega^*(M))^n$ *as* $\mathsf{D}\alpha = (d\alpha_1, d\alpha_2, \ldots, d\alpha_n)$, *where d denotes the exterior derivative.*

Definition 4.3.2 *Let M be a compact manifold and let n be a natural number We call an idempotent E in* $\mathrm{M}(n, C^\infty(M))$ a smooth idempotent *over M.*

Definition 4.3.3 *Let M be a compact manifold, let n be a natural number, and suppose that E is a smooth idempotent in* $\mathrm{M}(n, C^\infty(M))$. *The* Levi-Civita connection *on E is the map*

$$\mathsf{EDE} : (\Omega^*(M))^n \longrightarrow (\Omega^*(M))^n.$$

We write the Levi-Civita connection on E as ∇_E.

Our notion of Levi-Civita connection is an algebraic formulation of the Levi-Civita connection in differential geometry.

Definition 4.3.4 *Let E be a smooth idempotent over a compact manifold. Then $d\mathsf{E}$ is the matrix of one-forms obtained by applying the exterior derivative d to each entry of E.*

Do not confuse $d\mathsf{E}$ with the composition DE.

For each natural number n and each manifold M, there is a bilinear product

$$(\Omega^*(M))^n \times \Omega^*(M) \longrightarrow (\Omega^*(M))^n$$

defined by componentwise wedge product, and for $\alpha \in \left(\Omega^k(M)\right)^n$ and $\omega \in \Omega^l(M)$, two applications of the Leibniz rule give us

$$\begin{aligned}
\nabla_\mathsf{E}(\alpha\omega) &= \mathsf{EDE}(\alpha\omega) \\
&= \mathsf{E}(d\mathsf{E})(\alpha\omega) + \mathsf{E}(\mathsf{D}\alpha)\omega + (-1)^k \mathsf{E}\alpha d\omega \\
&= (\nabla_\mathsf{E}\alpha)\omega + (-1)^k \mathsf{E}\alpha d\omega.
\end{aligned}$$

Thus the Levi-Civita connection is not $\Omega^*(M)$-linear. However, because $d^2\omega = 0$ and $E\nabla_E = \nabla_E E$, we obtain

$$
\begin{aligned}
\nabla_E^2(\alpha\omega) &= \nabla_E\left((\nabla_E\alpha)\omega + (-1)^k E\alpha d\omega\right) \\
&= \left(\nabla_E^2\alpha\right)\omega + (-1)^{k+1}E(\nabla_E\alpha)d\omega + (-1)^k E\left(\nabla_E(\alpha\, d\omega)\right) \\
&= \left(\nabla_E^2\alpha\right)\omega + (-1)^{k+1}E(\nabla_E\alpha)d\omega + (-1)^k E(\nabla_E\alpha)d\omega + E\alpha\, d^2\omega \\
&= \left(\nabla_E^2\alpha\right)\omega,
\end{aligned}
$$

whence ∇_E^2 is $\Omega^*(M)$-linear.

Definition 4.3.5 *Let* E *be a smooth idempotent over a compact manifold* M. *The* curvature *of* E *is the* $\Omega^*(M)$-*linear operator* ∇_E^2.

We record two lemmas that are useful in computing curvature of an idempotent.

Lemma 4.3.6 *Let* M *be a compact manifold, let* n *be a natural number, and suppose that* E *in* $\mathrm{M}(n, C^\infty(M))$ *is a smooth idempotent. Then*

$$
\begin{aligned}
E(dE) &= (dE)(I_n - E) \\
(dE)E &= (I_n - E)(dE).
\end{aligned}
$$

Proof Take the exterior derivative on both sides of the equation $E^2 = E$ and apply the Leibniz rule on the left side to obtain

$$
(dE)E + E(dE) = dE.
$$

Then solve for $(dE)E$ and $E(dE)$. $\qquad\qquad\square$

Lemma 4.3.7 *Let* M *be a compact manifold, let* n *be a natural number, and suppose that* E *in* $\mathrm{M}(n, C^\infty(M))$ *is a smooth idempotent. Then*

$$
\nabla_E^2 = E(dE)^2 = (dE)(I_n - E)(dE) = (dE)^2 E.
$$

Proof For all α in $(\Omega^*(M))^n$, we compute

$$
\begin{aligned}
\nabla_E^2\alpha &= EDEDE\alpha \\
&= EDE\left((dE)\alpha + E(D\alpha)\right) \\
&= EDE(dE)\alpha + EDE(D\alpha) \\
&= E(dE)(dE)\alpha - E(dE)(D\alpha) + E(dE)(D\alpha) + E(D^2\alpha) \\
&= E(dE)^2\alpha.
\end{aligned}
$$

The rest of the lemma comes from applying Lemma 4.3.6 to the expression $E(dE)(dE)$. □

Note that in contrast to the scalar situation, the product of a matrix of one-forms with itself is not usually zero.

Proposition 4.3.8 (Bianchi identity) *Let M be a compact manifold and let ∇^2_E be the curvature of a smooth idempotent E over M. Then*

$$d\nabla^2_E = (dE)\nabla^2_E - \nabla^2_E(dE).$$

Proof From Lemmas 4.3.6 and 4.3.7, we obtain

$$(dE)\nabla^2_E = (dE)E(dE)^2 = (dE - E(dE))(dE)^2 = (dE)^3 - E(dE)^3,$$

and therefore

$$(dE)\nabla^2_E - \nabla^2_E(dE) = (dE)^3 - E(dE)^3 - E(dE)^3 = (dE)^3 = d\nabla^2_E.$$

□

Definition 4.3.9 *Let n be a natural number and suppose that P is an invariant polynomial on $M(n, \mathbb{C})$. For each smooth idempotent E in $M(n, C^\infty(M))$, define the differential form $P(\nabla^2_E)$ by formally evaluating P at ∇^2_E.*

Theorem 4.3.10 (Chern-Weil) *Let M be a compact manifold, let n be a natural number, and suppose that E in $M(n, C^\infty(M))$ is a smooth idempotent. Then for every invariant polynomial P on $M(n, \mathbb{C})$:*

(i) *the differential form $P(\nabla^2_E)$ is closed;*

(ii) *the cohomology class of $P(\nabla^2_E)$ in $H^*_{deR}(M)$ depends only on $[E]$ in $\mathrm{Idem}(C(X))$.*

(iii) *for $f : N \longrightarrow M$ smooth, we have $P(f^*(E)) = f^*(P(E))$ in $H_{deR}(N)$.*

Proof The sum and wedge of closed forms are closed, so to prove (i), it suffices by Corollaries 4.2.11 and 4.2.12 to verify that $\mathrm{Tr}\left(\nabla^{2k}_E\right)$ is closed for each k. We have the string of equalities

$$
\begin{aligned}
d\left(\mathrm{Tr}\left(\nabla_{\mathsf{E}}^{2k}\right)\right) &= \mathrm{Tr}\left(d\left(\nabla_{\mathsf{E}}^{2k}\right)\right) \\
&= k\,\mathrm{Tr}\left(d(\nabla_{\mathsf{E}}^2)\,\nabla_{\mathsf{E}}^{2(k-1)}\right) \\
&= k\,\mathrm{Tr}\left((d\mathsf{E})\nabla_{\mathsf{E}}^2\nabla_{\mathsf{E}}^{2(k-1)} - \nabla_{\mathsf{E}}^2(d\mathsf{E})\nabla_{\mathsf{E}}^{2(k-1)}\right) \\
&= 0;
\end{aligned}
$$

the first line follows from the linearity of the exterior derivative, the second line is obtained by repeatedly applying the Liebniz rule and employing the cyclic invariance of the trace, the third line is the substitution of the Bianchi identity into our expression, and the fourth line is again a consequence of the cyclic property of the trace.

To prove (ii), first note that

$$
\mathrm{Tr}(\nabla_{\mathrm{diag}(\mathsf{E},0)}^{2k}) = \mathrm{Tr}(\nabla_{\mathsf{E}}^{2k})
$$

for every natural number k and every smooth idempotent E. This observation, paired with Corollaries 4.2.11 and 4.2.12, shows that for any invariant polynomial P on $\mathrm{M}(n,\mathbb{C})$, the cohomology class of $P(\nabla_{\mathsf{E}}^2)$ in $H_{deR}^*(M)$ does not depend on the size of matrix we use to represent E.

To complete the proof of (ii), we will show that if $\{\mathsf{E}_t\}$ is a smooth homotopy of idempotents, then the partial derivative of $P\left(\nabla_{\mathsf{E}_t}^2\right)$ with respect to t is an exact form. To simplify notation, we will suppress the t subscript, and we will denote the derivative of E with respect to t as $\dot{\mathsf{E}}$. The sum and wedge product of exact forms are exact, so from Corollary 4.2.12 we deduce that we need only show that the partial derivative of $b_k(\nabla_{\mathsf{E}}^2)$ with respect to t is exact for each k.

We begin by noting three facts. First, from the equation $\mathsf{E}^2 = \mathsf{E}$, we have $\dot{\mathsf{E}}\mathsf{E} + \mathsf{E}\dot{\mathsf{E}} = \dot{\mathsf{E}}$, and thus $\mathsf{E}\dot{\mathsf{E}} = \dot{\mathsf{E}}(I_n - \mathsf{E})$. Second, if we multiply this last equation by E, we see that $\mathsf{E}\dot{\mathsf{E}}\mathsf{E} = 0$. Third, because mixed partials are equal, the partial derivative of $d\mathsf{E}$ with respect to t is $d\dot{\mathsf{E}}$.

For each k, we have

$$
\begin{aligned}
\frac{\partial}{\partial t}b_k(\nabla_{\mathsf{E}}^2) &= \mathrm{Tr}\left((\mathsf{E}(d\mathsf{E})(d\mathsf{E}))^k\right) \\
&= \mathrm{Tr}\left(\mathsf{E}(d\mathsf{E})^{2k}\right) \\
&= \mathrm{Tr}\left(\dot{\mathsf{E}}(d\mathsf{E})^{2k}\right) + \sum_{j=1}^{2k}\mathrm{Tr}\left(\mathsf{E}(d\mathsf{E})^{j-1}(d\dot{\mathsf{E}})(d\mathsf{E})^{2k-j}\right),
\end{aligned}
$$

where the second line follows from k applications of Lemma 4.3.7. We also have

$$\text{Tr}\left(\dot{\text{E}}(d\text{E})^{2k}\right) = \text{Tr}\left(\left(\dot{\text{E}}\text{E} + \text{E}\dot{\text{E}}\right)(d\text{E})^{2k}\right)$$

$$= \text{Tr}\left(\dot{\text{E}}\text{E}(d\text{E})^{2k}\right) + \text{Tr}\left(\text{E}\dot{\text{E}}(d\text{E})^{2k}\right)$$

$$= \text{Tr}\left(\dot{\text{E}}\text{E}\text{E}(d\text{E})^{2k}\right) + \text{Tr}\left(\text{E}\text{E}\dot{\text{E}}(d\text{E})^{2k}\right)$$

$$= \text{Tr}\left(\dot{\text{E}}\text{E}(d\text{E})^{2k}\text{E}\right) + \text{Tr}\left(\text{E}\dot{\text{E}}(d\text{E})^{2k}\text{E}\right)$$

$$= \text{Tr}\left(\text{E}\dot{\text{E}}\text{E}(d\text{E})^{2k}\right) + \text{Tr}\left(\text{E}\dot{\text{E}}\text{E}(d\text{E})^{2k}\right)$$

$$= 0 + 0 = 0,$$

and thus the first term of the derivative vanishes. From the remaining terms, we obtain

$$\sum_{j=1}^{2k} \text{Tr}\left(\text{E}(d\text{E})^{j-1}(d\dot{\text{E}})(d\text{E})^{2k-j}\right) = \sum_{j=1}^{2k} \text{Tr}\left((d\text{E})^{2k-j}\text{E}(d\text{E})^{j-1}(d\dot{\text{E}})\right)$$

$$= \sum_{j\ \text{even}} \text{Tr}\left((d\text{E})^{2k-j}\text{E}(d\text{E})^{j-1}(d\dot{\text{E}})\right) + \sum_{j\ \text{odd}} \text{Tr}\left((d\text{E})^{2k-j}\text{E}(d\text{E})^{j-1}(d\dot{\text{E}})\right)$$

$$= \sum_{j\ \text{even}} \text{Tr}\left((d\text{E})^{2k-1}(I_n - \text{E})(d\dot{\text{E}})\right) + \sum_{j\ \text{odd}} \text{Tr}\left((d\text{E})^{2k-1}\text{E}(d\dot{\text{E}})\right)$$

$$= \sum_{j\ \text{even}} \text{Tr}\left((d\text{E})^{2k-1}(d\dot{\text{E}}) - (d\text{E})^{2k-1}\text{E}(d\dot{\text{E}})\right) + \sum_{j\ \text{odd}} \text{Tr}\left((d\text{E})^{2k-1}\text{E}(d\dot{\text{E}})\right)$$

$$= k\,\text{Tr}\left((d\text{E})^{2k-1}(d\dot{\text{E}})\right)$$

$$= d\left(k\,\text{Tr}\left((d\text{E})^{2k-1}\dot{\text{E}}\right)\right).$$

Therefore $\frac{\partial}{\partial t}b_k(\nabla_\text{E}^2)$ is exact.

Finally, to establish (iii), note that the exterior derivative commutes with pullbacks, which implies that $\nabla_{f^*\text{E}} = f^*(\nabla_\text{E})$. Moreover, the functorial properties of f^* give us $P(\nabla_{f^*\text{E}}^2) = f^*(P(\nabla_\text{E}^2))$ for every invariant polynomial, and thus $P(f^*\text{E}) = f^*(P(\text{E}))$. $\qquad\square$

An important consequence of Theorems 1.7.14 and 4.3.10 is that for every invariant polynomial P, the cohomology class of $P(\nabla_\text{E})$ depends only the (smooth) isomorphism class of the vector bundle $\text{Ran}\,\text{E}$. We can therefore make the following definition.

Definition 4.3.11 *Suppose that V is a smooth vector bundle over a compact manifold M and let P be an invariant polynomial. We define* *the de Rham cohomology class $[P(V)]$ to be the de Rham cohomology class of $P(\nabla_E)$ for any smooth idempotent E with the property that V is isomorphic to $\operatorname{Ran} \mathsf{E}$.*

Definition 4.3.12 *Let V be a smooth vector bundle over a compact manifold M and suppose that P is an invariant polynomial. The de Rham cohomology class of $P(V)$ is called a* characteristic class *of V, and the integral of $P(V)$ over M is called a* characteristic number. *The de Rham cohomology classes $c_1(V)$, $c_2(V)$, $\ldots$, $c_m(V)$ are called the* Chern classes *of V, and their integrals over M are called* Chern numbers.

Theorem 4.3.13 *Let V be a smooth vector bundle over a compact manifold M and suppose that P is an invariant polynomial. Then $P(V)$ is a polynomial in the Chern classes of V.*

Proof Follows immediately from Theorems 4.2.9 and 4.3.10. $\qquad\square$

Example 4.3.14 *Let V be the vector bundle on S^2 associated to the smooth idempotent*

$$\mathsf{E} = \frac{1}{2} \begin{pmatrix} 1+x & y+iz \\ y-iz & 1-x \end{pmatrix}$$

from Example 1.4.3. We compute

$$d\mathsf{E} = \frac{1}{2} \begin{pmatrix} dx & dy+i\,dz \\ dy-i\,dz & -dx \end{pmatrix}$$

and

$$(d\mathsf{E})(d\mathsf{E}) = \frac{1}{2} \begin{pmatrix} -i\,dy\,dz & dx\,dy+i\,dx\,dz \\ -dx\,dy+i\,dx\,dz & i\,dy\,dz \end{pmatrix}$$

Thus

$$\nabla_{\mathsf{E}}^2 = \mathsf{E}(d\mathsf{E})(d\mathsf{E}) = \frac{1}{4} \begin{pmatrix} \alpha_{11} & \alpha_{12} \\ \alpha_{21} & \alpha_{22} \end{pmatrix},$$

where α_{11}, α_{12}, α_{21}, and α_{22} are two-forms. Therefore

$$\det\left(1+\nabla_{\mathsf{E}}^2\right) = \det \begin{pmatrix} 1+\frac{1}{4}\alpha_{11} & \frac{1}{4}\alpha_{12} \\ \frac{1}{4}\alpha_{21} & 1+\frac{1}{4}\alpha_{22} \end{pmatrix}$$

$$= \left(1+\frac{1}{4}\alpha_{11}\right)\left(1+\frac{1}{4}\alpha_{22}\right) - \frac{1}{16}\alpha_{12}\alpha_{21}$$

$$= 1 + \frac{1}{4}(\alpha_{11}+\alpha_{22}),$$

because all four-forms on S^2 are zero. We also know in this case that

$$\det\left(1 + \nabla_E^2\right) = 1 + c_1(\nabla_E^2).$$

Comparing the two expressions and doing an easy computation yields

$$c_1(\nabla_E^2) = -\frac{i}{2}(\alpha_{11} + \alpha_{22})$$

$$= -\frac{i}{2}\left(z\,dx\,dy - y\,dx\,dz + x\,dy\,dz\right).$$

We will show that $c_1(\nabla_E^2)$ determines a nontrivial de Rham cohomology class by showing that the first Chern number is not zero. This is most easily accomplished by converting to spherical coordinates:

$$\int_{S^2} c_1(\nabla_E^2) = -\frac{i}{2}\int_{S^2} z\,dx\,dy + y\,dx\,dz - x\,dy\,dz$$

$$= \frac{i}{2}\int_0^\pi\int_0^{2\pi}\left(\sin\phi\cos^2\phi + \sin^2\theta\sin^3\phi + \cos^2\theta\sin^3\phi\right)\,d\theta\,d\phi$$

$$= 2\pi i.$$

Thus we have shown that V is not trivial.

The ease with which we worked out the previous example may leave the reader with a false impression of the computations typically involved. If a vector bundle appears from some topological construction, it may be difficult to realize it as the range of an idempotent. In addition, the forms representing the Chern classes of bundles that are not line bundles can be vastly more complicated. For this reason, we often look for indirect ways to compute Chern classes and numbers (see the exercises).

There are several characteristic classes that arise naturally in geometry and topology. To describe them, let

$$f(z) = a_0 + a_1 z + a_2 z^2 + \cdots + a_k z^k + \cdots$$

be a power series with complex coefficients. For each matrix A in $M(m, \mathbb{C})$, define

$$\Pi_f(A) = \det(f(A)).$$

We are only taking wedge products of forms of even degree, so wedge product is commutative and the determinant is well defined. Note that Π_f is an invariant polynomial for each f.

Definition 4.3.15 *Let V be a smooth vector bundle over a compact manifold M, let $f(z)$ be a complex formal power series, and choose a* *smooth idempotent with the property that* $V \cong \operatorname{Ran} \mathsf{E}$. *The* multiplicative (characteristic) class $\Pi_f(V)$ *is the cohomology class of* $\det\left(f(\nabla_{\mathsf{E}}^2)\right)$ *in* $H_{deR}^{even}(M)$.

Note that the convergence of the power series f is not important here because $\nabla_{\mathsf{E}}^{2k} = 0$ for $2k$ greater than the dimension of M.

Example 4.3.16 *The multiplicative class associated to the polynomial* $1 + z$ *is called the* total Chern class; *for a smooth vector bundle, this class is usually writen $c(V)$. If we write out this characteristic class, we obtain*

$$c(V) = 1 + c_1(V) + c_2(V) + \cdots + c_k(V) + \cdots.$$

Example 4.3.17 *The multiplicative class associated to*

$$\frac{z}{1 - e^{-z}} = 1 + \frac{1}{2}z + \frac{1}{12}z^2 + \cdots$$

is called the Todd class. *The Todd class measures the way in which the Thom isomorphism in K-theory differs from its analogue in de Rham cohomology.*

Example 4.3.18 *The multiplicative class associated to*

$$\frac{\sqrt{z}}{\tanh\sqrt{z}} = 1 + \frac{1}{3}z - \frac{1}{45}z^2 + \cdots$$

is called the L-class. *The L-class is intimately connected to a topological invariant of a manifold called the* signature.

Example 4.3.19 *The multiplicative class associated to*

$$\frac{\frac{\sqrt{z}}{2}}{\sinh\left(\frac{\sqrt{z}}{2}\right)} = 1 - \frac{1}{24}z + \frac{7}{5760}z^2 + \cdots$$

is called the $\widehat{A}$-class. *The $\widehat{A}$-class arises in the study of spin structures and manifolds that admit Riemannian metrics with positive scalar curvature.*

The next proposition justifies calling the characteristic classes "multiplicative."

Proposition 4.3.20 *Let V and W be smooth vector bundles over a compact manifold and let $f(z)$ be a formal power series. Then*

$$\Pi_f(V \oplus W) = \Pi_f(V)\Pi_f(W),$$

where the product on the right-hand side is induced by the wedge product.

Proof Choose smooth projections E and F whose ranges are V and W respectively. Then $V \oplus W$ is isomorphic to $\mathrm{Ran}(\mathrm{diag}(\mathsf{E}, \mathsf{F})$, and

$$\nabla_{\mathrm{diag}(\mathsf{E},\mathsf{F})} = \mathrm{diag}(\nabla_\mathsf{E}, \nabla_\mathsf{F}).$$

The proposition then follows from the definition of Π_f and properties of the determinant. $\square$

4.4 The Chern character

In the last section, we used formal power series and the determinant to define multiplicative characteristic classes. By replacing the determinant with the trace, we obtain a new collection of characteristic classes.

Definition 4.4.1 *Let V be a smooth vector bundle over a compact manifold M, let $f(z)$ be a complex formal power series, and choose a smooth idempotent with the property that $V \cong \mathrm{Ran}\,\mathsf{E}$. The additive (characteristic) class $\Sigma_f(V)$ is the cohomology class of $\mathrm{Tr}\left(f(\nabla_\mathsf{E}^2)\right)$ in $H_{deR}^{even}(M)$.*

Proposition 4.4.2 *Let V and W be smooth vector bundles over a compact manifold M and let $f(z)$ be a formal power series. Then*

$$\Sigma_f(V \oplus W) = \Sigma_f(V) + \Sigma_f(W).$$

Proof The proof of this proposition is the same as that of Proposition 4.3.20, with the multiplicative property of the determinant replaced by the additive property of the trace. $\square$

Corollary 4.4.3 *Let $f(z)$ be a formal power series and let M be a compact manifold. Then the additive class Σ_f determines a group homomorphism (which we denote by the same symbol)*

$$\Sigma_f : \mathrm{K}^0(M) \longrightarrow H_{deR}^{even}(M)$$

with the property that

$$\Sigma_f([\mathsf{E}] - [\mathsf{F}]) = \Sigma_f(\mathsf{E}) - \Sigma_f(\mathsf{F})$$

for all smooth idempotents E *and* F *over* M.

Proof Follows immediately from Proposition 4.4.2 and Theorem 1.6.7.

$\square$

Definition 4.4.4 *The* Chern character *is the additive class associated to the power series*

$$e^z = 1 + z + \frac{z^2}{2} + \cdots .$$

The Chern character of a smooth vector bundle V *is denoted* $\mathrm{Ch}(V)$.

There is a corresponding "odd" Chern character from $\mathrm{K}^{-1}(M)$ to $H_{deR}^{odd}(C^\infty(M))$. As we did for $\mathrm{K}^0(M)$, we will accept without proof that every element of $\mathrm{K}^{-1}(M)$ can be represented by an invertible matrix with smooth entries and that if two such matrices can be connected by a continuous path, then they can also be connected by a smooth path.

Definition 4.4.5 *Let* M *be a compact manifold and let* n *be a natural number. For each* S *in* $\mathrm{GL}(n, C^\infty(M))$, *define*

$$\widetilde{\mathrm{Ch}}(\mathsf{S}) = \sum_{k=0}^{\infty} (-1)^k \frac{k!}{(2k+1)!} \, \mathrm{Tr}\left(\left(\mathsf{S}^{-1} d\mathsf{S} \right)^{2k+1} \right).$$

Lemma 4.4.6 *Let* M *be a compact manifold and suppose that* W *is a matrix of one-forms on* M. *Then* $\mathrm{Tr}(\mathsf{W}^2) = 0$.

Proof Denote the (i, j) entry of W as ω_{ij}. Then

$$\mathrm{Tr}(\mathsf{W}^2) = \sum_{i,j} \omega_{ij}\omega_{ji} = \sum_{i<j} \omega_{ij}\omega_{ji} + \sum_{i>j} \omega_{ij}\omega_{ji}$$

$$= \sum_{i<j} \omega_{ij}\omega_{ji} + \sum_{i<j} \omega_{ji}\omega_{ij} = \sum_{i<j} \omega_{ij}\omega_{ji} - \sum_{i<j} \omega_{ij}\omega_{ji} = 0.$$

$\square$

Lemma 4.4.7 *Let* M *be a compact manifold, let* n *be a natural number, and suppose that* S *is an element of* $\mathrm{GL}(n, C^\infty(M))$. *Then*

$$d(\mathsf{S}^{-1}) = -\mathsf{S}^{-1}(d\mathsf{S})\mathsf{S}^{-1}.$$

Proof Apply the exterior derivative to both sides of the equation $S^{-1}S = I_n$ and solve for $d(S^{-1})$. $\qquad\square$

Lemma 4.4.8 *Let M be a compact manifold, let n be a natural number, and suppose that S is an element of $\mathrm{GL}(n, C^\infty(M))$. Then*

$$d\big((S^{-1}dS)^{2k}\big) = 0$$

for all natural numbers k.

Proof Lemma 4.4.8 and the Leibniz rule imply that

$$
\begin{aligned}
d\big((S^{-1}dS)^2\big) &= d(S^{-1}dS)(S^{-1}dS) - (S^{-1}dS)d(S^{-1}dS) \\
&= -S^{-1}(dS)S^{-1}(dS)S^{-1}(dS) + S^{-1}(dS)S^{-1}(dS)S^{-1}(dS) \\
&= 0.
\end{aligned}
$$

The lemma then follows by induction. $\qquad\square$

Proof We need only verify the formula for $k = 1$; the general result then follows from the Leibniz rule and mathematical induction. From Lemma 4.4.7 we obtain

$$
\begin{aligned}
d\big((S^{-1}dS)^2\big) &= d(S^{-1}dS)(S^{-1}dS) - (S^{-1}dS)d(S^{-1}dS) \\
&= -S^{-1}(dS)S^{-1}(dS)S^{-1}(dS) + S^{-1}(dS)S^{-1}(dS)S^{-1}(dS) = 0.
\end{aligned}
$$

$\qquad\square$

Theorem 4.4.9 *Let M be a compact manifold and let n be a natural number. For every S in $\mathrm{GL}(n, C^\infty(M))$, the form $\widetilde{\mathrm{Ch}}(S)$ is closed and its cohomology class depends only on $[S]$ in $\mathrm{K}^{-1}(M)$.*

Proof For each natural number k, we have

$$
\begin{aligned}
d\,\mathrm{Tr}\big((S^{-1}dS)^{2k+1}\big) &= \mathrm{Tr}\big(d\,((S^{-1}dS)^{2k+1})\big) \\
&= \mathrm{Tr}\big(d(S^{-1}dS)(S^{-1}dS)^{2k} - (S^{-1}dS)d\,((S^{-1}dS)^{2k})\big) \\
&= -\,\mathrm{Tr}\big(S^{-1}(dS)S^{-1}(dS)(S^{-1}dS)^{2k}\big) \\
&= -\,\mathrm{Tr}\big((S^{-1}dS)^{2k+2}\big).
\end{aligned}
$$

This quantity vanishes by Lemma 4.4.8, and thus we see from Definition 4.4.5 that $\widetilde{\mathrm{Ch}}(S)$ is closed. To establish the rest of the theorem, first note that $\widetilde{\mathrm{Ch}}(\mathrm{diag}(S, 1)) = \widetilde{\mathrm{Ch}}(S)$, so the size of the matrix we use to represent S in $\mathrm{GL}(C^\infty(M))$ does not matter. Suppose that $[S_0] = [S_1]$

in $K^{-1}(M)$ for some S_0 and S_1 in $GL(n, C^\infty(M))$. Let $\{S_t\}$ be a smooth homotopy of invertibles from S_0 to S_1. We will show that the partial derivative of $Ch(S_t)$ with respect to t is an exact form. As in our proof of Theorem 4.3.10, we will simplify notation by suppressing the t subscript, and we will denote the derivative of S with respect to t by $\dot{S}$. For each natural number k, we have the equalities

$$\frac{\partial}{\partial t} \operatorname{Tr} \left(S^{-1}dS\right)^{2k+1} = \operatorname{Tr} \sum_{j=0}^{2k} \left(S^{-1}dS\right)^j \left(\frac{\partial}{\partial t}\left(S^{-1}dS\right)\right) \left(S^{-1}dS\right)^{2k-j}$$

$$= \operatorname{Tr} \sum_{j=0}^{2k} \left(\frac{\partial}{\partial t}\left(S^{-1}dS\right)\right) \left(S^{-1}dS\right)^{2k}$$

$$= (2k+1)\operatorname{Tr}\left(\frac{\partial}{\partial t}\left(S^{-1}dS\right)\right)\left(S^{-1}dS\right)^{2k}$$

$$= (2k+1)\operatorname{Tr}\left(\left(-S^{-1}\dot{S}S^{-1}dS + S^{-1}d\dot{S}\right)\left(S^{-1}dS\right)^{2k}\right)$$

$$= (2k+1)\operatorname{Tr}\left(-S^{-1}\dot{S}\left(S^{-1}dS\right)^{2k+1}\right)$$

$$\qquad + (2k+1)\operatorname{Tr}\left(S^{-1}d\dot{S}\left(S^{-1}dS\right)^{2k}\right),$$

where we have used the cyclic property of the trace to go from the first line to the second. This form is exact, because

$$d\left((2k+1)\operatorname{Tr}\left(S^{-1}\dot{S}\left(S^{-1}dS\right)^{2k}\right)\right)$$

$$= (2k+1)\operatorname{Tr}\left(-S^{-1}(dS)S^{-1}\dot{S}\left(S^{-1}dS\right)^{2k} + S^{-1}d\dot{S}\left(S^{-1}dS\right)^{2k}\right)$$

$$= (2k+1)\operatorname{Tr}\left(-S^{-1}\dot{S}\left(S^{-1}dS\right)^{2k+1}\right) + (2k+1)\operatorname{Tr}\left(S^{-1}d\dot{S}\left(S^{-1}dS\right)^{2k}\right);$$

we have again used the cyclic property of the trace and Lemma 4.4.8. Thus

$$\operatorname{Tr}\left(\left(S^{-1}dS\right)^{2k+1}\right)$$

is exact for each k, and the desired result is then a consequence of the definition of $\widetilde{Ch}(S)$. $\qquad\square$

Definition 4.4.10 *Let M be a compact manifold and let n be a natural number. For each S in $GL(n, C^\infty(M))$, define the Chern character of S to be the class $Ch(S)$ of $\widetilde{Ch}(S)$ in $H^{odd}_{deR}(M)$.*

Proposition 4.4.11 *For every compact manifold M, the Chern character is a group homomorphism from $K^{-1}(M)$ to $H^{odd}_{deR}(M)$.*

Proof Corollary 2.3.7 implies that

$$[\mathsf{ST}] = [\operatorname{diag}(\mathsf{S}, I)][\operatorname{diag}(I, \mathsf{T})] = [\operatorname{diag}(\mathsf{S}, \mathsf{T})]$$

in $\mathrm{K}^{-1}(M)$ for all invertible matrices S and T over M. Thus we obtain

$$\begin{aligned}
\operatorname{Tr}\Big(\big(\operatorname{diag}(\mathsf{S}, \mathsf{T})d(\operatorname{diag}(\mathsf{S}, \mathsf{T}))\big)^{2k+1}\Big) \\
= \operatorname{Tr}\big(\operatorname{diag}((\mathsf{S}^{-1}d\mathsf{S})^{2k+1}, (\mathsf{T}^{-1}d\mathsf{T})^{2k+1})\big) \\
= \operatorname{Tr}\big((\mathsf{S}^{-1}d\mathsf{S})^{2k+1}\big) + \operatorname{Tr}\big((\mathsf{T}^{-1}d\mathsf{T})^{2k+1}\big)
\end{aligned}$$

for each natural number k; Definitions 4.4.5 and 4.4.10 then imply that Ch is a group homomorphism (note the group operation on $\mathrm{K}^{-1}(M)$ is written multiplicatively, while on $H_{deR}^*(M)$, addition is the group operation). $\qquad\square$

Example 4.4.12 *Take S^1 to be the unit circle in $\mathbb{C}$, and suppose that $f \in C^\infty(S^1)$ is nowhere vanishing. Then $\mathrm{Ch}(f)$ is defined and is the cohomology class of $f^{-1}\,df$. By definition, the integral of the differential form $f^{-1}\,df$ over S^1 is $2\pi i$ times an integer called the* winding number *of f. Roughly speaking, the winding number of f counts the number of times the image of f "wraps around" the origin counterclockwise.*

Corollary 4.4.3 and Proposition 4.4.11 show that for every compact manifold M, we can combine the "even" and "odd" Chern characters combine to obtain a group homomorphism

$$\mathrm{Ch} : \mathrm{K}^0(M) \oplus \mathrm{K}^{-1}(M) \longrightarrow H_{deR}^*(M);$$

see the notes for more information about this map.

4.5 Notes

The material in this chapter on differential forms and de Rham cohomology is adapted from [6]; see that book for more informations. It is difficult to find a good source that discusses smoothness of vector bundles, idempotents, and invertibles, but some of this material can be found in Section 3.8 of [9].

Many authors define multiplicative and additive characteristic classes as scalar multiples of ours, with the multiple usually being $(2\pi i)^{-1}$. This modification allows us to consider Chern classes and the Chern character as classes in Čech (or singular) cohomology with integer coefficients. Because we only work with de Rham cohomology in this book, we forego the factor of $(2\pi i)^{-1}$. The extra work in defining the Chern character on a compact manifold M so that it takes values in the integer cohomology groups $H^*(M, \mathbb{Z})$ has the payoff that $\mathrm{Ch} : \mathrm{K}^*(M) \longrightarrow H^*(M, \mathbb{Z})$ is a group isomorphism, and, in fact, is even a ring isomorphism. It is possible to define characteristic classes for vector bundles over any compact Hausdorff space, and in this situation, the Chern character becomes an isomorphism if we ignore torsion elements; i.e., elements of finite order. For the details of these facts about the Chern character, the reader should consult [13].

Exercises

4.1   Show that the vector bundle over the torus whose range is the idempotent in Example 1.4.5 is nontrivial.

4.2   For each natural number k, compute the Chern classes of vector bundle V that is the range of the idempotent

$$\frac{1}{|z_1|^{2k} + |z_2|^{2k}} \begin{pmatrix} |z_1|^{2k} & z_1 \bar{z}_2^k \\ \bar{z}_1^k z_2^k & |z_2|^{2k} \end{pmatrix}$$

over $\mathbb{CP}^1$.

4.3   We say vector bundles V and W over a compact Hausdorff space X are *stably isomorphic* if $V \oplus \Theta^k(X) \cong W \oplus \Theta^l(X)$ for some natural numbers k and l. Prove that if V and W are smooth stably isomorphic vector bundles over a compact manifold, then $c_i(V) = c_i(W)$ for all natural numbers i.

4.4   Let V be the vector bundle over S^2 that is the range of the idempotent

$$\begin{pmatrix} x^2 & xy & xz \\ xy & y^2 & yz \\ xz & yz & z^2 \end{pmatrix}.$$

Compute the Chern classes of V.

4.5   Let V be a smooth vector bundle over a 5-dimensional compact manifold M. Show that the Todd class of V can be written

$$\frac{1}{2}c_1(V) + \frac{1}{12}\left(c_1^2(V) + c_2(V)\right) + \frac{1}{24}c_1(V)c_2(V).$$

4.6   Let V and W be smooth vector bundles over a compact manifold and suppose that $V \oplus W$ is trivial. Show that

$$c_1(W) = -c_1(V)$$
$$c_2(W) = c_1^2(V) - c_2(V)$$
$$c_3(W) = 2c_1(V)c_2(V) - c_3(V) - c_1^3(V).$$

4.7   Let V be a smooth vector bundle over a compact manifold and suppose that $V \oplus V$ is trivial. What can you say about the Chern classes of V?

4.8   Let V be a bundle over a compact manifold M, let W be a subbundle of V, and let Q denote the quotient bundle (see Exercise 6). Prove that $c(V) = c(W)c(Q)$.

4.9   Suppose V is a rank n vector bundle over a compact manifold. Prove that if V has a nowhere vanishing smooth section, then $c_n(V) = 0$.

4.10   Let n be a natural number and suppose that S is an element of $\mathrm{GL}(n, C^\infty(S^1))$. Show that

$$\int_{S^1} \widetilde{Ch}(S) = 2\pi i(\text{winding number of det } S).$$

References

[1] Adams, J.F. (1960). On the non-existence of elements of Hopf invariant one, *Ann. of Math.* **72** (2), 20–104.

[2] Adams, J.F. and Atiyah, M.F. (1966). *K*-theory and the Hopf invariant, *Quart. J. Math. Oxford Ser.* **17** (2), 31–38.

[3] Atiyah, M.F. (1967). *K-theory*, New York: Benjamin.

[4] Atiyah, M.F. and Hirzebruch, F. (1961). Vector bundles and homogemeous spaces, *Proc. Sympos. Pure Math., vol III*, 7–38, Providence: Amer. Math. Soc.

[5] Blackadar, B. (1998). *K-theory for Operator Algebras*, 2nd ed, MSRI Publication Series no. 5, New York: Springer-Verlag.

[6] Bott, R. and Tu, L.V. (1982). *Differential Forms in Algebraic Topology*, Graduate Texts in Mathematics, vol. 82, New York: Springer-Verlag.

[7] Dupont, J. (1968) *K-Theory*, Lecture Notes Series, no. 11, Math. Inst., Aarhus Univ., Aarhus.

[8] Eilenberg, S. and Steenrod, N. (1952). *Foundations of Algebraic Topology*, Princeton: Princeton University Press.

[9] Gracia-Bondía, J., Várilly, J., and Figueroa, H. (2001) *Elements of Noncommutative Geometry*, Boston: Birkhäuser.

[10] Hungerford, T. (1980). *Algebra*, Graduate Texts in Mathematics, vol. 73, New York: Springer-Verlag.

[11] Husemoller, D. (1975). *Fiber Bundles*, 2nd ed, Graduate Texts in Mathematics, vol. 20, New York: Springer-Verlag.

[12] Karoubi, M. (1967). *Séminaire Cartan-Schwartz 1963/64, exposé 16*, New York: Benjamin.

[13] Karoubi, M. (1978). *K-theory: An introduction*, Grundlehren der Math. Wissenschaften, vol. 226, New York: Springer-Verlag.

[14] Karoubi, M. (1980). *Some Applications of Topological K-theory*, Notas de Matemática, no. 74, Amsterdam: North-Holland.

[15] Rosenberg, J. (1994). *Algebraic K-theory and Its Applications*, 2nd ed, Graduate Texts in Mathematics, vol. 147, New York: Springer-Verlag.

[16] Wegge-Olsen, N.E. (1993). *K-theory and C*-algebras. A Friendly Approach*, New York: Oxford University Press.

Symbol index

Subject index